AF245586

THE PUPIL

With an Introduction by

Irving H. Leopold, M.D.
*Professor and Chairman
Department of Ophthalmology
The Mount Sinai Hospital
New York, New York*

and Forewords by

Henry F. Allen, M.D.
*Henry Willard Williams Professor of Ophthalmology
Chairman, Ophthalmology Department
Harvard Medical School
Cambridge, Massachusetts*

and

Frank B. Walsh, M.D.

*Professor Emeritus
Department of Ophthalmology
The Wilmer Ophthalmological Institute
The Johns Hopkins University School of Medicine
Baltimore, Maryland*

THE PUPIL

By

KEITH M. ZINN, M.D.
*Instructor
Department of Ophthalmology
Mount Sinai School of Medicine
City University of New York
New York, New York*

CHARLES C THOMAS • PUBLISHER
Springfield • Illinois • U.S.A.

Published and Distributed Throughout the World by

CHARLES C THOMAS • PUBLISHER

BANNERSTONE HOUSE

301-327 East Lawrence Avenue, Springfield, Illinois, U.S.A.

This book is protected by copyright. No part of it may be reproduced in any manner without written permission from the publisher.

© 1972, by CHARLES C THOMAS • PUBLISHER

ISBN 0-398-02320-4

Library of Congress Catalog Card Number: 79-187682

With THOMAS BOOKS *careful attention is given to all details of manufacturing and design. It is the Publisher's desire to present books that are satisfactory as to their physical qualities and artistic possibilities and appropriate for their particular use.* THOMAS BOOKS *will be true to those laws of quality that assure a good name and good will.*

Printed in the United States of America

W-2

To

my mother and father
who have always encouraged me
in the pursuit of knowledge

FOREWORD

THE MATERIAL IN this monograph was first presented by its author to his classmates in the Harvard Basic Science Course in Ophthalmology. Their response was enthusiastic. In subsequent years he has returned to both the Harvard and Lancaster courses as a faculty member. Again his ratings by the students have been of the highest. If every student were to work up a subject as Dr. Zinn has done with *The Pupil,* there would be no need for faculty, and students could teach each other. The work of the course director would be simplified and students could evaluate each other's performance.

Since this utopia is not about to come into being, the next best thing is widespread availability of the course outline. Dr. Zinn's superb organization of his material has resulted in a text that will have widespread appeal to students of all ages—neurologists as well as ophthalmologists. Its predictable success will be the measure of the need it will meet.

HENRY F. ALLEN

FOREWORD

THE INVITATION TO write a foreword to this manual is an honor
that is deeply appreciated.

In his preface, Dr. Zinn remarks on what really amounted to
very little help from me; also he graciously acknowledges information obtained from *Clinical Neuro-ophthalmology* by Walsh
and Hoyt. I hasten to remark that Dr. William F. Hoyt and
Dr. H. Stanley Thompson in major part are responsible for any
real value Chapter 4 may have; it deals with the autonomic
system, the pupil, accommodation and tear secretion.

What Dr. Zinn has produced is a manual that will be useful
to clinicians. The anatomy and the physiology, as it is known,
are described clearly. The figures which are included in the text
are excellent and clearly labelled, and greatly aid in understanding what in many instances is complex. In correlating basic
information with symptomatology, one soon becomes aware that
fact and fancy are not confused.

In recommending this publication without reservation, I do so
with a firm belief that a manual concerning a particular feature
of a subject—any subject—is a useful contribution. Here the
clinical evaluation of pupillary states is vitally important in
evaluations of central nervous system disorders.

Finally, what Dr. Zinn has produced in this manual may
well serve as evidence of what can be accomplished when ability,
with persistence and sound judgment, are centered on an
objective.

FRANK B. WALSH

PREFACE

This manual has been written for ophthalmologists, neurologists, neurosurgeons and other students of neuropathology in an attempt to clarify the various pupillary reactions that occur under a myriad of clinical conditions. The structure of the iris as well as the neuro-anatomical pathways that influence the intrinsic musculature of the iris are reviewed in detail. In addition, the various pharmacological agents and physiological reflexes that are part and parcel of pupillary behavior are discussed. In the last portion of the manual, space is devoted to the correlation of the above concepts with the clinical features of the pupil under various pathological states.

I am deeply indebted to Professor Frank B. Walsh for allowing me to read "The Autonomic Nervous System: The Pupil, Accommodation and Tear Secretion," which is a chapter in the new edition of *Clinical Neuro-Ophthalmology,* edited by Walsh and Hoyt. Many of the didactic statements in this manual, covering the more controversial aspects of neuro-ophthalmology, reflect the experience and opinions of Professor Walsh as expressed in his chapter on the pupil.

Also, I wish to thank Doctor Edward Wolpow, Assistant Professor of Neurology, The Massachusetts General Hospital, for his invaluable comments and criticisms of the manuscript, as well as Professor R. J. Last, of London, England, for his help with the various aspects of the neuroanatomical pathways.

The author gratefully acknowledges the technical assistance of Miss Nancy Vander Pyl, who did a remarkable job in preparing the illustrations for this manual.

Lastly, I am most indebted to Professor Henry F. Allen and Doctor B. Thomas Hutchinson of the Harvard Medical School for their encouragement and advice, without which this book could not have been written.

Keith Marshall Zinn

INTRODUCTION

An understanding of the anatomy, physiology, pharmacology and clinical attitudes of the pupil provides a strong foundation for all ophthalmologists.

Many disciplines of medicine are brought together in this structure.

It is not an easy task to present the essential material concerning such an important area. Dr. Zinn has delivered many lectures on this subject over the past several years. He has worked in this area experimentally and clinically. His writings reveal a thorough grasp. All the essential pupillary data are here. This text can serve as an introduction to the novice, stimulating review for the busy practitioner, and as a refreshing presentation for anyone seeking rapid reference in these specific areas. Time may be our most precious possession. Dr. Zinn has respected this by making the text clear, concise, and complete but brief.

Irving H. Leopold

CONTENTS

THE PUPIL

Chapter 1

ANATOMY OF THE IRIS

THE PUPIL, WHEN functioning normally, is linked with a servo-mechanism within the central nervous system that controls the amount of light energy traversing the lens of the eye and reaching the retinal receptors. The size of the pupil also influences the depth of focus as well as the amount of chromatic and spherical aberration of the optical system within the eye. In addition to these important optical functions, the pupil serves as an important neurological indicator in certain disease states of the central and peripheral nervous systems. It is the purpose of this monograph to delineate the normal neuroanatomical and neurophysiological relationships (circuits and reflexes) that control pupillary size and function and demonstrate how these dynamic relationships are altered in various disease states. Let us begin with the gross anatomy of the iris as described below.

GROSS ANATOMY

The iris is the most anterior portion of the vascular tunic of the eye. At its periphery, the iris root is attached to the middle of the anterior surface of the ciliary body. The axial border, which is unattached, forms a central (actually it is slightly nasoinferior to the true geometric center of the iris) aperture termed the *pupil*. Under normal conditions, the pupillary margin of the iris rests upon and is supported by the anterior surface of the lens (lens capsule) which is located in a plane more anterior to that of the plane formed by the origin of the iris from the ciliary body. Therefore, the iris inclines anteriorly as one goes from its attached or ciliary margin to its free or pupillary margin. The resultant three-dimensional shape of the iris is similar to that of a truncated cone (a cone whose apex is cut off forming the pupillary border). When the lens is absent or dislocated posteriorly secondary to trauma, the cone is flattened and the iris is noted to be tremulous. This condition or pattern of iris movement is known as *iridodonesis*. Although the pupillary margin of the iris rests on the anterior surface of the lens, it is lifted anteriorly, from time to time, by the aqueous humor percolating through from the posterior chamber to the anterior chamber.

The thinnest portion of the iris is found at the peripheral root or ciliary margin, while the thickest portion of the iris is found at the collarette. Because of the extreme thinness of the iris root, it is often the site of iris damage in contusion injuries of the eye. In cases where the iris root separates from the ciliary body, usually secondary to ocular trauma, there is an open space between the posterior and anterior chambers peripherally which is termed *iridodialysis*.

In general, the anterior surface of the iris is irregular (iris crypts, contraction furrows, collarette etc.), while the posterior surface is relatively smooth. The pupil varies in diameter according to the actions of two sets of intrinsic muscles, namely the *sphincter pupillae* and the *dilator pupillae*. The relationship of these muscles to the rest of the iris tissue will be found in a schematized drawing of the iris in cross section (see Fig. 1). Finally, the iris can be considered as a thin membranous diaphragm located between the cornea and lens serving to compartmentalize the anterior segment of the eye into anterior and posterior chambers.

Anterior Surface Structure of the Iris

When one examines the anterior surface of the iris either with a loupe or with a slit lamp, one cannot fail to be impressed by its marked irregularity. Approximately 1.5 mm from the pupillary margin there is an irregular line which is somewhat scalloped in appearance that is termed the *collarette* or *iris frill*. The collarette is formed by the underlying *circulus vasculosus iridis minor* ("circulus arteriosus iridis minor" has been used to denote this structure, but it is a poor term because the vascular complex is an arteriovenous anastomosis). The prominence of

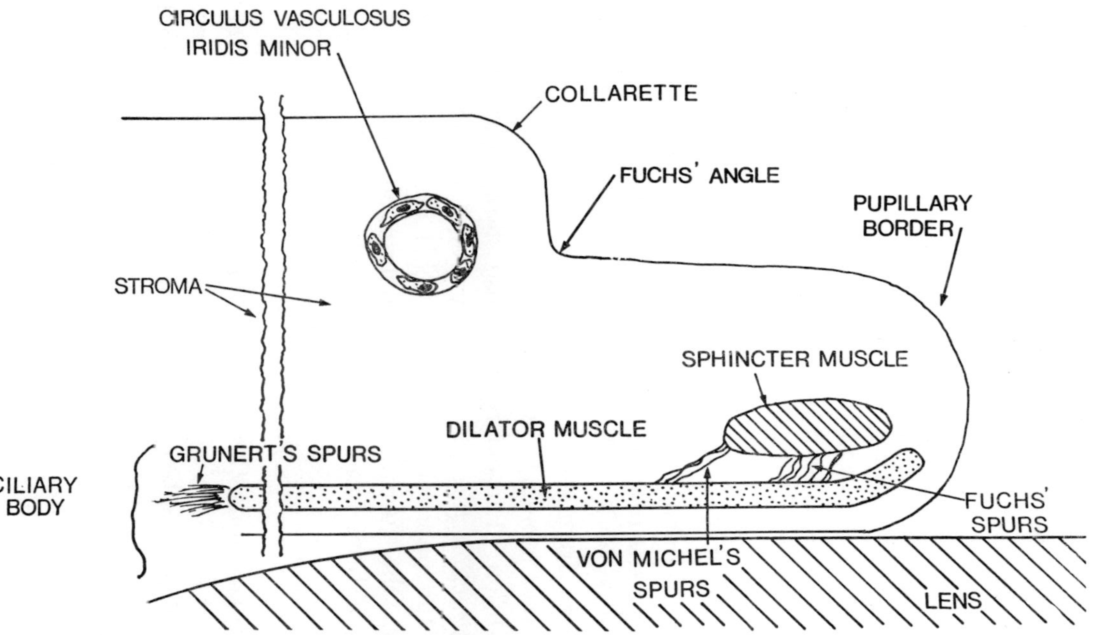

Figure 1. Cross-section of iris. Note the relationship of the sphincter muscle to the dilator muscle. The collarette is located on the anterior iris surface. Grunert's, Von Michel's and Fuchs' spurs are believed to be part of the dilator pupillae.

the collarette is enhanced by the fact that the anterior vascular layer of the iris is absent in the zone central to the collarette (pupillary zone). The collarette is also the point at which the pupillary membrane is attached during embryonic development. Lastly, the collarette divides the iris into two zones: the ciliary zone and the pupillary zone.

Ciliary Zone

The ciliary zone (greater ring of Merkel) is located peripheral to the collarette and is wider than the pupillary zone. In the region of the circulus vasculosus minor there are numerous pit-like depressions called the *crypts of Fuchs*. These appear to be channels through which aqueous fluid can pass during contraction and dilation of the pupil. The ciliary zone also shows a radial structural quality which is due to the radial arrangement of iris blood vessels running from the major arterial circle located within the ciliary body to the minor vascular circle located within the collarette. In cases of iris atrophy, the radial structural quality is rather prominent and may be the most significant sign in this condition. The ciliary zone also shows concentric lines termed *contraction furrows*. These are most prominent during dilation of the pupil and consist of folds of iris tissue whose valleys form the furrows. These contraction rings are most conspicuous in brown irides and least conspicuous in blue irides.

Pupillary Zone

The pupillary zone (lesser ring of Merkel) is located central to the collarette and its width changes as the pupillary aperture is altered. The pupillary zone is widest when the pupillary opening is smallest. As the pupil dilates, however, the width of the pupillary zone of the iris decreases. In fact, in early dilation, increase in pupillary diameter occurs at the expense of the pupillary zone which decreases in width. In the fully dilated pupil, the pupillary zone of iris tissue rolls up under the collarette (telescopic motion) and is barely visible when viewing the iris in the frontal plane.

During embryonic development, the pupillary zone is filled

with mesoderm which forms the *pupillary membrane* at 4-months gestation (110 mm stage). At 8-months gestation (230-265 mm stage) the pupillary membrane begins to regress. At birth, however, many infants can be seen to have remnants of the pupillary membrane. In adults, remnants of this membrane are often observed on slit-lamp examination. On occasion, the membrane fails to regress at all and this condition is termed *persistent pupillary membrane*. In this case, strands of iris tissue are seen emanating from the collarette and not the pupillary border as in synechiae secondary to inflammation or uveitis. Neither the collarette nor the pupillary crypts are present at birth.

As one traverses the pupillary zone of iris on the anterior surface and approaches the collarette, there is a sharp step anteriorly which is termed *Fuchs' angle*. Also, in some cases, one can see the outline of the sphincter pupillae as a circular pale white band about 1 mm in width adjacent to the pupillary margin (especially on slit-lamp examination of blue irides).

Puff or Pigment Seam of the Iris

The puff or pigment seam of the iris is located at the pupillary margin where there is a thin rim of darkly pigmented tissue easily seen on slit-lamp examination that represents the most anterior portion of the optic cup. It is formed by the eversion of the two posterior pigmented cell layers of the iris. As the pupil dilates, the width of this seam increases and is called *physiological ectropion uvea*. In pathological states, the pigmented seam of tissue may be pulled toward the periphery to a higher degree than normal, becoming visible to the naked eye, and is termed a *pathological ectropion uvea*. There is associated fibrovascular tissue that is responsible for the pathological uveal ectropion.

Pupillary Margin

The pupillary margin, on close examination, is not round and smooth, but consists of folds or crenations (about 70 in number). It is reasoned that these folds allow for a change in the circumference of the pupil during dilation and contraction. These folds become prominent in the miotic state (small pupil).

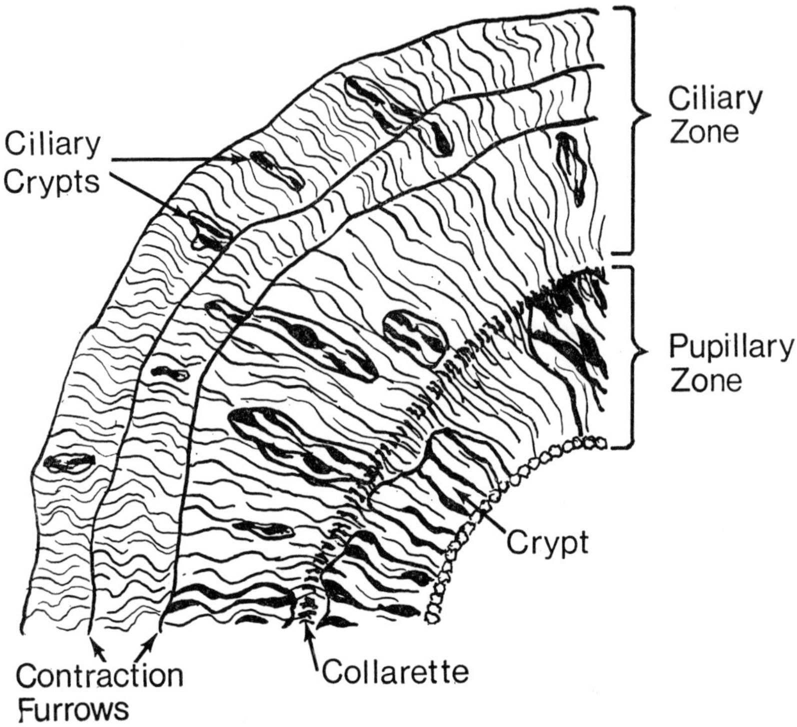

Figure 2.   Surface anatomy of the iris.   Note that the ciliary and pupillary zones of the iris are divided by the collarette.   The contraction furrows are formed by rolls of iris tissue during dilation of the pupil.   (From Walsh, F. B., and Hoyt, W. F.: *Clinical Neuro-Ophthalmology*, 3rd ed. Baltimore, Williams & Wilkins, 1969.)

Posterior Surface Structure of the Iris

In general, this surface is darker and smoother than the anterior surface of the iris.   There are *contraction folds* (first system of Schwalbe) that are radial, each one being about 1 mm in length, located about the pupillary margin.   There are 70 of these folds and they are deeper in the miotic state.   At 1.5 mm from the pupillary margin, directly underneath the collarette, there are *structural folds* (second system of Schwalbe) which are radial in direction and extend toward the periphery.   They are formed by the radial vessels and are continuous with the valleys between the ciliary processes. In the periphery, there are *circular furrows* similar to those on the anterior surface.

Blood Vessels of the Iris

The long posterior ciliary and anterior ciliary arteries merge within the ciliary body to form the *circulus vasculosus iridis major.* From this arterial vascular complex, radial vessels run axially toward the region of the collarette where a second vascular complex is formed, namely the *circulus vasculosus iridis minor.* The minor vascular circle is not arterial in nature but rather an arteriovenous anastomosis. From this complex, fine vessels run to the pupillary margin and nourish the sphincter pupillae. The veins of the iris transmit blood back to the vortex veins.

The blood vessels comprise the bulk of the iris stroma and can be found coursing through the stroma in a sinuous, yet radial, fashion. The coiled state of these vessels allows easy adaptation to rapidly changing states of contraction of iris musculature (mydriasis, miosis). Histologically, iris arteries have a thick adventitia and a thin muscular coat. Iris veins have perivascular connective tissue sheaths. The overall iris circulation is shown in Figure 3.

Innervation of the Iris

There are five components to the innervational complex of the iris, which are as follows:

1. Sensory modalities conducted to the central nervous system through the nasociliary branch of the Vth cranial nerve (first, or ophthalmic, division).
2. Sympathetic vasomotor efferents which supply the blood vessels of the ciliary body and iris. These are postganglionic fibers that travel on the internal carotid artery, pass through the ciliary ganglion without synapsing, and terminate on the vascular complex of the ciliary body and iris.
3. Postganglionic sympathetic efferent fibers that travel the same route as the sympathetic vasomotor fibers, as outlined above, but terminate on the chromatophores (branched) of the iris.
4. Sympathetic pupillomotor fibers that are postganglionic

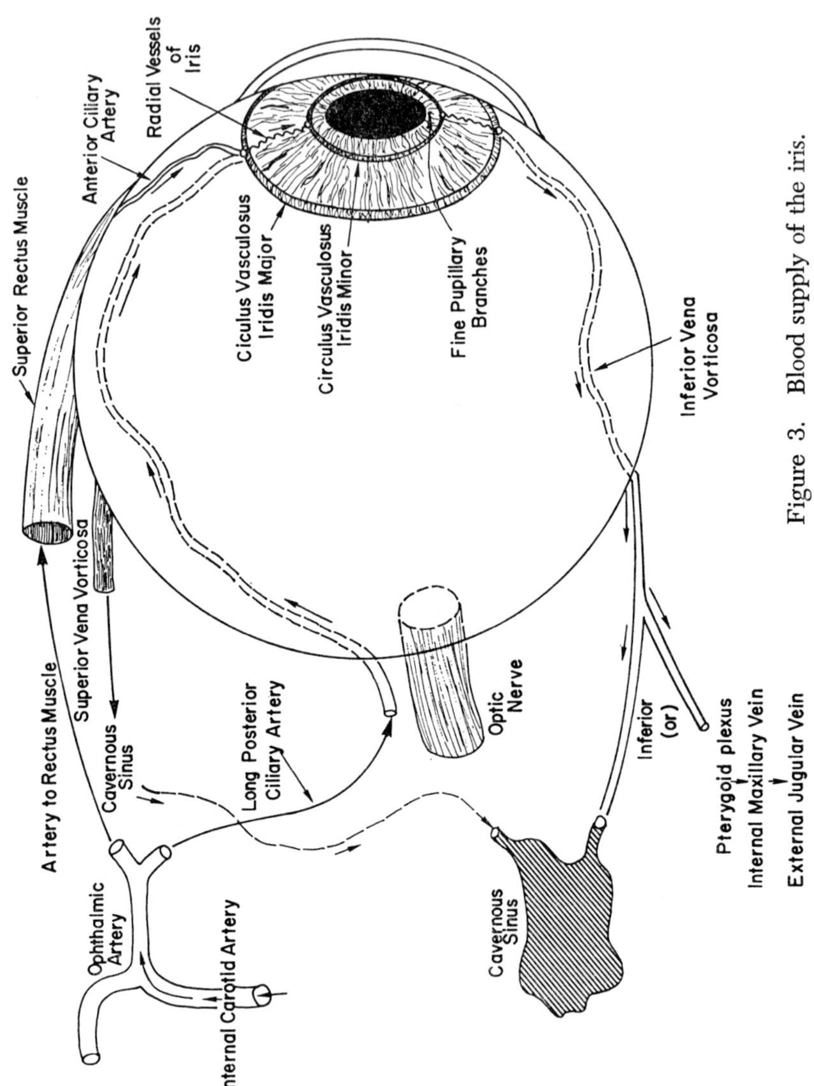

Figure 3.   Blood supply of the iris.

terminate upon the dilator pupillae. The preganglionic fibers originate in the ciliospinal center of Budge which is located in the spinal cord (C_8-T_2).

5. Postganglionic parasympathetic fibers that originate in the ciliary ganglion and terminate upon the sphincter pupillae. Preganglionic fibers originate in the Edinger-Westphal nucleus within the mesencephalon.

The neuroanatomical pathways that these nerves traverse will be considered in greater detail in later sections of the manual under the heading of "Neuroanatomical Pathways of Iris Innervation."

MICROSCOPIC ANATOMY

With the exception of the pupillary zone, the iris consists of five layers, which are as follows:
1. The anterior endothelium.
2. The anterior limiting (border) layer.
3. The stroma or vascular layer.
4. The posterior membrane of Bruch.
5. The posterior epithelium.

Anterior Endothelium

There is considerable controversy among various authors over the existence of this layer on the anterior surface of the iris. In fact, several authors have shown that within the iris crypts there is no anterior endothelial layer present. This finding implies that the aqueous has direct access to the iris stroma.

Anterior Limiting (Border) Layer

This is a narrow zone which represents the anterior condensation of the vascular layer. The anterior border layer is comprised of collagenous fibers, as well as branching melanocytes containing pigment. The anterior border layer is poorly developed in persons with blue irides yet is well developed in patients with brown irides. In the latter case, the pigmented melanocytes are profuse and form a relatively thick border layer. Obliterated blood vessels and numerous nerve terminals are found within this layer.

Within the pupillary zone and the ciliary zone, local deficiencies in the anterior border layer are termed *pupillary crypts* and *ciliary crypts*, respectively. Also, it is of note that at the contraction furrows, there is a deficiency of the anterior border layer.

Stroma or Vascular Layer

The stroma consists of loose connective tissue containing melanocytes which are embedded around the sphincter pupillae muscle and the blood vessels and nerves of the iris. It is interesting to note that the connective tissue fibers of the iris are mostly collagenous. The few elastic fibers which are present are found coursing, in a radial fashion or direction in the posterior portion of the stroma. In addition there are elastic fibers between the sphincter and dilator muscles. With regard to the cellular population within the iris stroma, one can find the usual assortment of fibroblasts, macrophages as well as plasma cells (see Fig. 4).

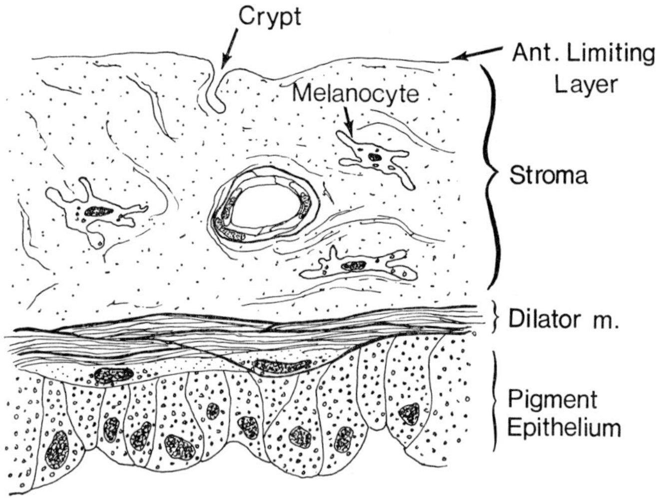

Figure 4. Cross-section of the iris. Note the double layered pigment epithelium posteriorly and its relationship to the dilator muscle fibers. This illustration is based on observations of bleached histological preparations of normal iris tissue (after Wolff). From Walsh, F. B., and Hoyt, W. F.: *Clinical Neuro-Ophthalomology,* 3rd ed. Baltimore, Williams & Wilkins, 1969.

The pigmented melanocytes or chromatophores of the iris are of two types. *Branching melanocytes* have numerous long, slender cell processes filled with yellow and brown pigment granules. These cells are found in the anterior and posterior layers of the iris as well as around the blood vessels, ciliary margin and sphincter muscle. Postganglionic sympathetic fibers innervate these cells. *Clump cells* are ovoid in shape and do not have the long, slender cell processes of the branching melanocytes. Clump cells contain dark pigment granules and are found throughout the vessel layer in deeply pigmented irides. There is no sympathetic innervation of the clump cells and they are derived from the pigment epithelium embryologically. Clump cells are found predominantly in the pupillary zone near the sphincter muscle. However, the predominant type of pigmented cell found within the iris stroma is the branching melanocyte.

Sphincter Pupillae

The sphincter muscle consists of a flat bar of muscle tissue whose fibers run, for the most part, in a circular direction. It is approximately 1 mm wide, forming a continuous ring about the pupillary margin near the posterior surface of the iris. The sphincter is bound to the surrounding structures by connective tissue fibers and blood vessels. When the sphincter contracts (miosis), it tends to pull the edge of the iris pigment epithelium onto the anterior surface of the iris resulting in *physiological ectropion uvea* (see page 7). When the sphincter is cut, i.e. iridectomy, the remaining portion of the intact sphincter will respond to the light reflex with pupillary miosis. The sphincter muscle consists of 70 segments, each of which is innervated by a separate twig from the postganglionic parasympathetic outflow (see Fig. 5).

Embryologically, the sphincter is derived from the pigmented cells (ectodermal) located near the pupillary border. A comparison of the sphincter and dilator muscles can be found in Table I.

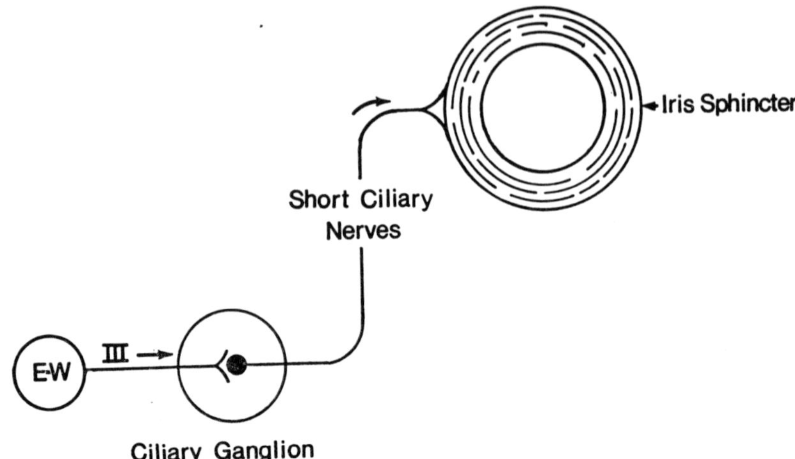

Figure 5. Iris sphincter. The Edinger-Westphal nucleus, located within the mesencephalon, sends parasympathetic pupillomotor fibers to the ciliary ganglion where they synapse with postganglionic fibers that eventually innervate the iris sphincter muscle.

TABLE I

COMPARISON OF SPHINCTER AND DILATOR MUSCLES

	Sphincter	*Dilator*
Origin	Ectodermal	Ectodermal
Development	Good at birth	Poor at birth
Mesodermal sheath	+	−
Pigmentation	−	+
Innervation	Parasympathetic Sympathetic	Sympathetic
	Segmental	Many overlapping sectors
Strength	More powerful	Less powerful
Change in length	Up to 87%	Large range, but less than the sphincter

Blood Vessels of the Iris

See page 9 for a description of blood vessels of the iris.

The Posterior Membrane

The posterior membrane of Bruch consists of a thin layer of smooth muscle fibers which, like the sphincter, are derived from the anterior layer of the optic cup and therefore are of ectodermal origin. These fibers constitute the *dilator pupillae muscle* and are part of the spindle-shaped cells of the iris pigment epithelium (see Fig. 6). The dilator muscle runs in a radial direction and

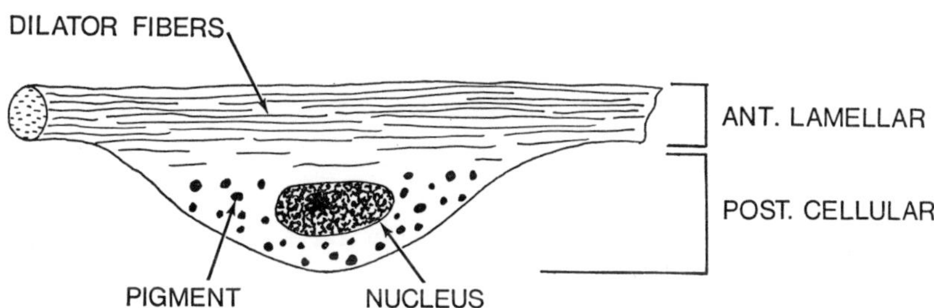

Figure 6. Iris dilator-muscle cell. Note that the dilator cell has a posterior portion which is pigmented and an anterior portion which contains the dilator-muscle fibers.

stops some 0.3 mm short of the pupillary border where the cells are entirely pigmented! Occasionally there are a few junctional fibers between the posterior mid-portion of the sphincter and the anterior surface of the dilator muscle, termed *Fuchs' spurs* (Fig. 1). Similar fibers connecting the dilator with the peripheral border of the sphincter are termed *von Michel's spurs* (Fig. 1). In both cases, the connective fibers have some pigment associated with them. At the iris root, there may be small bundles of dilator muscle extending toward the ciliary body which have been termed *Grunert's spurs* (Fig. 1). The dilator muscle continues peripherally, inserting into the ciliary body which is its site of origin. When the dilator pupil muscle constricts, the pupils dilate (mydriasis).

At birth, the dilator muscle is poorly developed, which explains the difficulty in obtaining good mydriasis with sympathomimetic agents. The dilator pupillae muscle is innervated by the sympathetic nervous system via the long ciliary nerves (Fig. 7). For a detailed discussion of the sympathetic innervation of the iris, the reader is referred to the section entitled "Neuroanatomical Pathways of Iris Innervation."

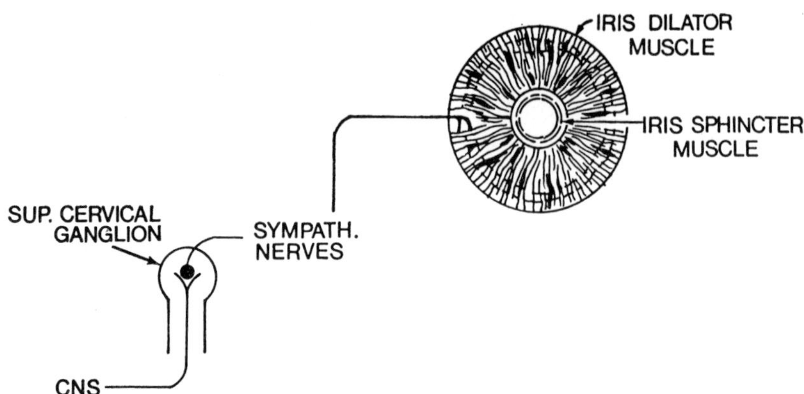

Figure 7. Iris dilator-muscle innervation. Note that the postganglionic sympathetic fibers originating in the superior cervical ganglion terminate upon the iris dilator muscle.

Posterior Epithelium

The iris pigment epithelium consists of two layers of cells (Fig. 4) which are derived from the anterior lip of the optic cup. They are densely pigmented and only in albinotic cases or in tissues that have been specifically bleached can one appreciate the true histological anatomy of these cells. The anterior layer consists of pigmented spindle-shaped cells (Fig. 6) with long, oval nuclei. The major axis of this cell is oriented in a radial direction. The posterior half of the cell contains the nucleus and pigment granules, while the anterior portion consists of dilator muscle fibers. Therefore, the pigmented spindle cells are *myoepithelial* in nature, in that one portion is serving as a pigmented epithelial layer and the other is functioning as a thin sheet of muscle.

The dilator muscle is innervated by postganglionic sympathetics and therefore sensitive to adrenergic agents. The pharmacologic as well as the neuroanatomic features of this sympathetic innervation will be covered in later sections of the manual.

NEUROANATOMICAL PATHWAYS OF IRIS INNERVATION
Afferent Sympathetic Nervous System Pathways

Retinohypothalamic projection fibers have been demonstrated in animals, but their existence in man is open to question. These fibers, when present, run from the retina through the optic nerve and exit in the posterior portion of the optic chiasm on their way to the hypothalamus. It is believed by some authors that these retinohypothalamic projections are involved in photoneuro-endocrine reflexes or photoperiodic functions which have been termed *circadian cycles*.

Efferent Sympathetic Nervous System Pathway

The hypothalamus is a rather complex structure with multiple association pathways. However, for purposes of describing its generalized functions, it can be divided into anterior and posterior portions. The *posterior hypothalamus* appears to be essential for pressor functions related to sympathetic nervous system activity. The *anterior hypothalamus* is associated with parasympathetic nervous system activity. Stimulation of the posterior hypo-thalamus results in heat production, increased basal metabolic rate, vasoconstriction and pupillary dilatation (see Fig. 30).

From the posterior and lateral regions of the hypothalamus, descending sympathetic (pupillary) efferent fibers enter the *prerubral field* and are distributed in the *capsule of the red nucleus* (see Fig. 8). Some of these fibers undergo a partial crossing of the midline within the *decussation of Forel*. However, the majority of the descending fibers are ipsilateral and continue to run toward a group of cells located within the grey matter, the *intermediolateral column* or *ciliospinal center of Budge and Waller*, of the spinal cord (C_8, T_{1-2}). It is probable that some of the descending fibers undergo synapse in the region of the mesencephalon and pons, but the majority go directly, without synapse, to the ciliospinal center via the *anterolateral columns*. When the ciliospinal center of Budge is stimulated, there are vasopressor responses as well as mydriasis indicating that pupillary and vasomotor modalities may have similar neuroanatomical pathways.

The second neuron (preganglionic) in this pathway leaves the ciliospinal center of Budge via the *ventral root* (C_8, T_{1-2}) to the *paravertebral sympathetic chain* using the *white rami com-municantes* (see Fig. 9). These fibers continue, without synapse,

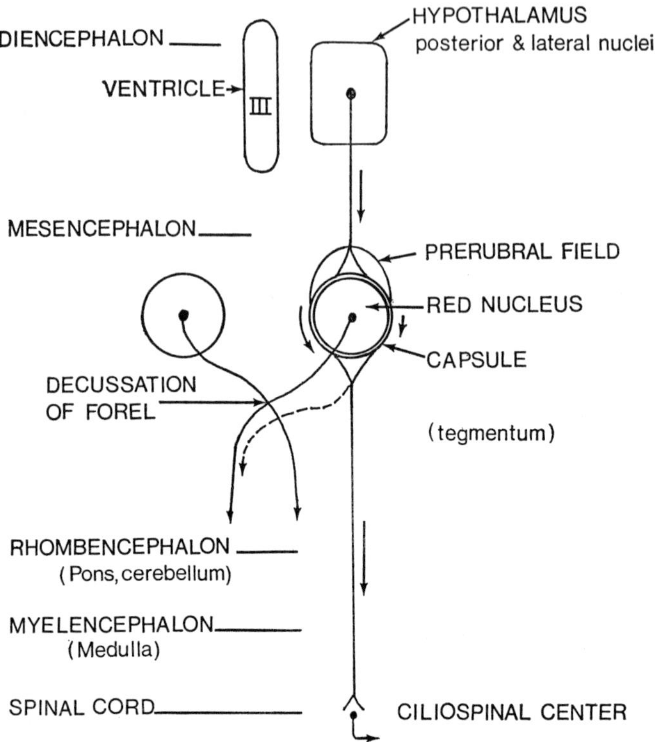

Figure 8. Sympathetic pathway, hypothalamic and mesencephalon levels. Note the schematic representation of the descending sympathetic fibers originating in the posterior and lateral regions of the hypothalamus. These fibers are found in the capsule of the red nucleus and a small percentage of the fibers may cross the midline within the decussation of Forel.

through the stellate ganglion, passing very close to the surface of the apical lung pleura (see Fig. 10). The majority of the fibers travel in the anterior loop of the *ansa subclavia* only to return to the main trunk of the sympathetic chain and subsequently ascend through the *inferior* and *middle cervical ganglia*. These fibers terminate in the *superior cervical ganglion.*

The superior cervical ganglion is often the largest sympathetic ganglion in the body. It is located at the base of the skull between the internal carotid artery and the jugular vein. Estimates of the preganglionic to postganglionic fiber ratio vary from 1:11 to 1:17.

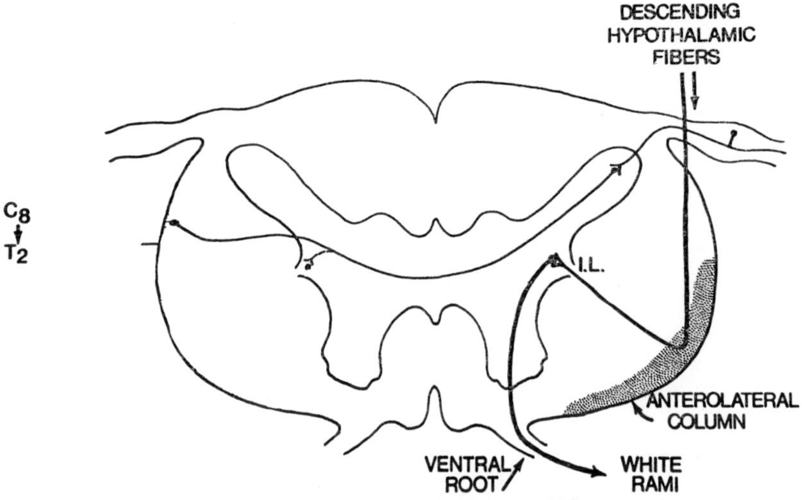

Figure 9. Sympathetic pathway, spinal cord level. Note the descending hypothalamic fibers within the anterolateral columns (shaded area) of the spinal cord, synapsing in the interomediolateral, *IL* or ciliospinal center of Budge and exiting the cord via the ventral root.

The third neuron (postganglionic fibers) in this sympathetic outflow leaves the superior cervical ganglion to run onto the common carotid artery and the internal carotid artery as a plexus. Some of these sympathetic pupillomotor fibers leave the internal carotid artery plexus and enter the middle ear where they join the *tympanic branch of the glossopharyngeal nerve (IX)* to form the *tympanic plexus* situated on the promonotory of the middle ear (see Fig. 11). These fibers are collectively termed the *caroticotympanic nerves.* When these fibers are damaged, as in an acute otitis media, an ipsilateral *Horner's syndrome* (see Chapter 6) may be present. The pupillary components of the tympanic plexus exit the middle ear and rejoin the internal carotid artery plexus, which in turn enters the *cavernous sinus.* Within the cavernous sinus, an arborization of the sympathetic fibers takes place, and the entire complex is termed the *cavernous sinus plexus.* It is at this juncture that most of the pupillary efferent fibers join the ophthalmic division of the trigeminal (V) nerve (see Fig. 12) and go on to run in the *nasociliary branch*

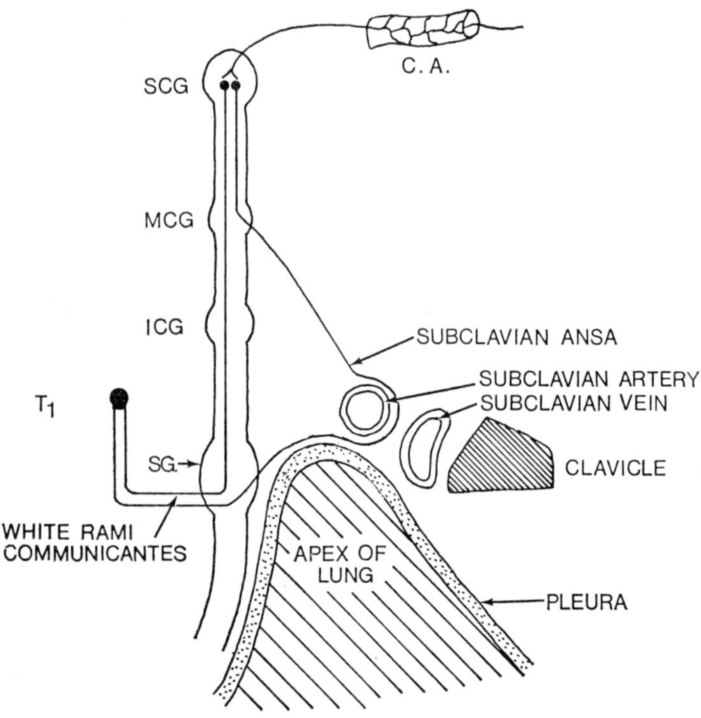

Figure 10. Sympathetic pathway, paravertebral sympathetic chain. Note that the majority of preganglionic sympathetic fibers exit the spinal cord at the T_1 level and enter the stellate ganglion, *SG*, via the white rami communicantes and then travel via the anterior loop of the ansa subclavia to the middle cervical ganglion, *MCG* and synapse in the superior cervical ganglion, *SCG*. Postganglionic fibers form a plexus about the carotid artery. In addition, note the close anatomical relationship of the ansa subclavia to the apical lung pleura.

and finally enter the globe via the *long ciliary nerves*. Once these fibers are within the globe, they run in the suprachoroidal space toward the ciliary body and then turn axially to innervate the dilator pupillae muscle.

The postganglionic sympathetic vasomotor fibers of the eye travel along the common carotid and internal carotid arteries as part of the carotid plexus. They enter the cavernous sinus and contribute to the cavernous sinus plexus; then these fibers

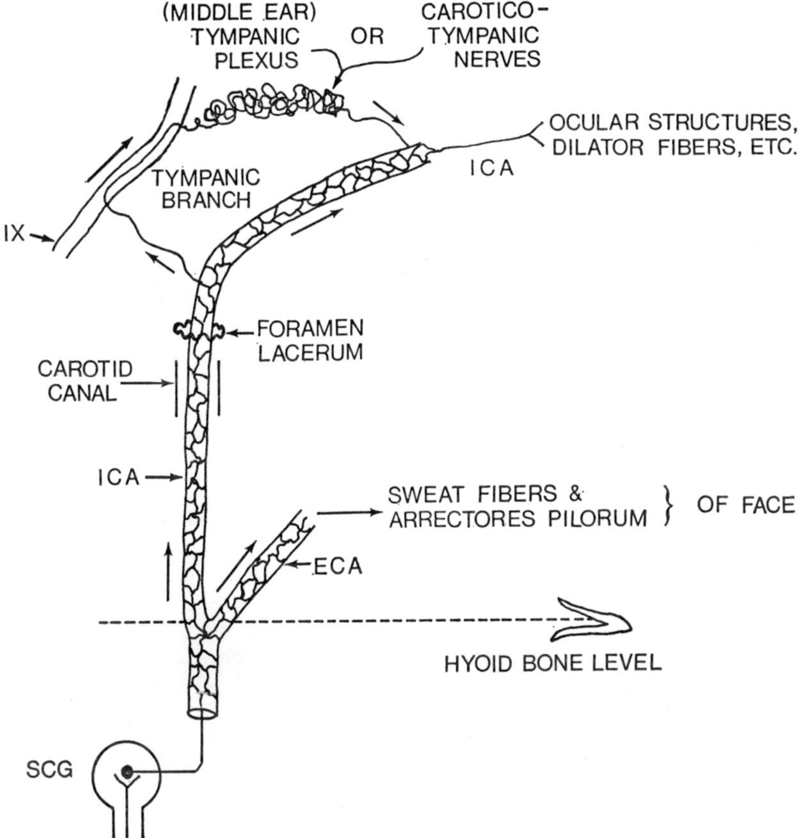

Figure 11. Sympathetic pathway, caroticotympanic nerves. Note that the internal carotid artery, *ICA*, plexus contributes fibers to the tympanic plexus in the middle ear. The majority of sympathetic fibers to the sweat glands and arrectores pilorum of the face travel with the external carotid artery, *ECA*.

enter the orbit via the *superior orbital fissure,* traveling on various blood vessels such as the ophthalmic artery. The vasomotor fibers leave the blood vessels and pass through the ciliary ganglion via the *sympathetic root* without synapsing and continue toward the globe in the *short ciliary nerves* (see Fig. 12) supplying blood vessels and branched chromatophores of the uveal tract (see Fig. 13). The sympathetics derived from the internal carotid

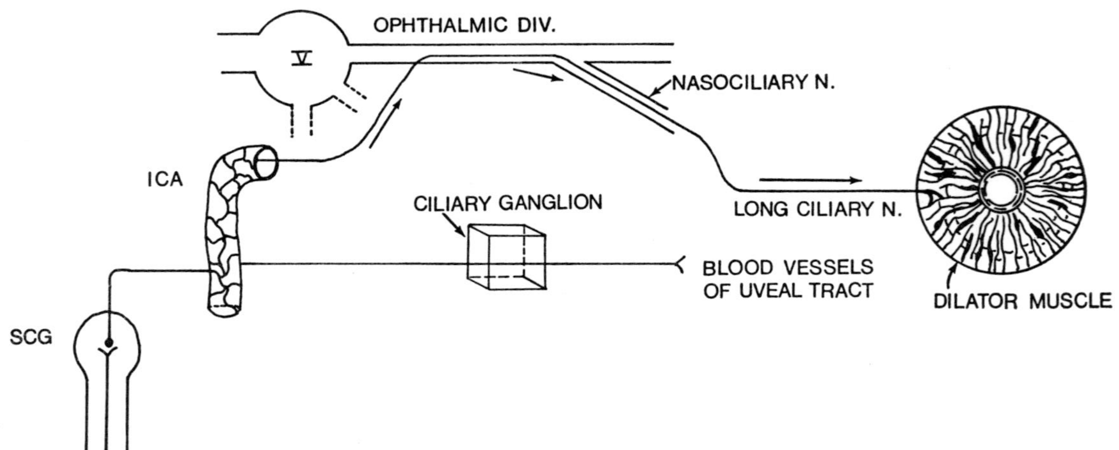

Figure 12. Sympathetic pathway, nasociliary and long ciliary nerves. Note that postganglionic pupillomotor fibers from the internal carotid artery, *ICA*, enter the ophthalmic division of the trigeminal nerve, *V*, and then go to the nasociliary branch and finally enter the globe via the long ciliary nerves to synapse on the dilator pupillae muscle.

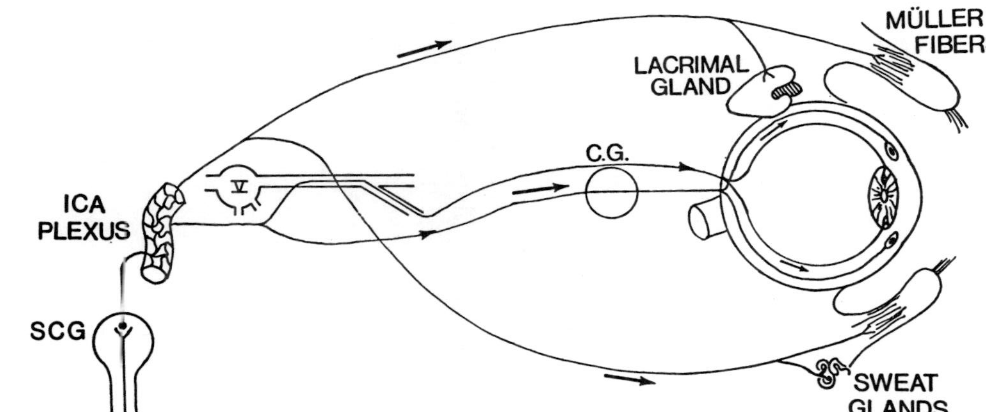

Figure 13. Sympathetic pathway to the ocular and intraorbital structures. *CG*, ciliary ganglion; SCG, superior cervical ganglion. Sympathetic vasomotor fibers from the internal carotid artery plexus, ICA, travel through the ciliary ganglion to innervate the blood vessels and branched melanocytes of the uveal tract.

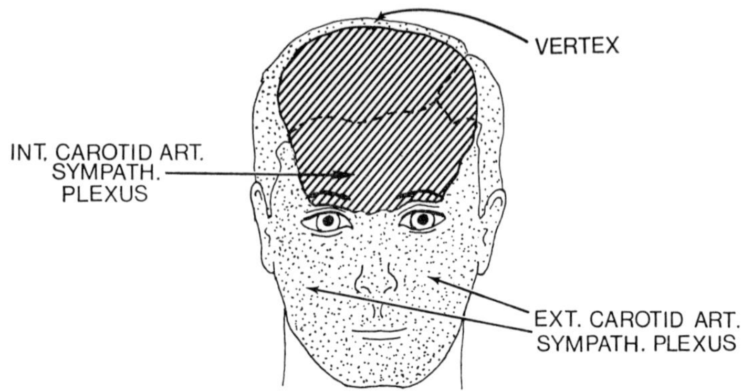

Figure 14.   Sympathetic supply to the facial area. Note that the arrectores pilorum and the sweat glands of the skin in between the brows and in the mid-forehead region are supplied by the sympathetics derived from the internal carotid artery plexus. The remainder of the skin in the facial area is supplied by the external carotid artery plexus.

artery plexus also supply the lacrimal gland and the smooth-muscle fibers of Muller located in the upper and lower eyelids as well as in the floor of the orbit (fibers covering part of the inferior orbital fissure). In addition, the sympathetics derived from the internal carotid artery plexus supply the *arrectores pilorum* and *sweat glands* in the skin of the face over the nasion and mid-forehead regions whereas the sympathetic fibers derived from the external carotid artery plexus supply these structures in the skin over the remaining portions of the face (see Figs. 11 and 14).

Afferent Parasympathetic Nervous System Pathways

Excitatory Pathways to the Edinger-Westphal Nucleus

Retinomesencephalic Pathway

The afferent arc of the light reflex commences with the stimulation of the retinal receptor cells (rods and cones) by quanta of light (400 to 700 mμ). Within the rod or cone cell there is a conversion of this light energy into chemical energy which in turn is relayed to a bipolar cell through a synaptic junction located in the outer plexiform layer (see Fig. 15). From

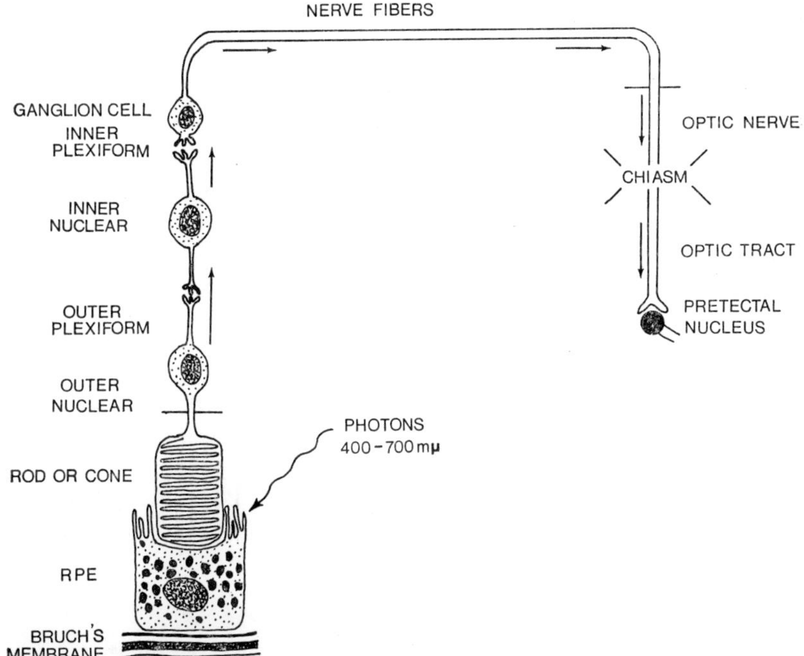

Figure 15. Schematic drawing of retinal pathway in the light reflex. Note that the neuronal connections as depicted in this illustration are highly schematic and represent an oversimplification of this process. However, the general sequence of events is believed to be as follows: photons of light, in the range of 400 to 700 mμ, strike the retinal receptor, i.e. rod or cone cell, where there is a complex series of photochemical reactions beginning with rhodopsin (11-cis form). The result of these reactions is a transmission of information from the receptor cell to possibly a bipolar cell across a synaptic junction known as the outer plexiform layer. A second possible relay is from the bipolar cell to the ganglion cell across a second synaptic junction known as the inner plexiform layer. From here, visual information is conducted to the brain via the optic nerve, optic chiasm and optic tract with synapse either in the lateral geniculate body (visual fibers) or in the pretectal region of the mesencephalon (afferent pupillary fibers). RPE is the retinal pigment epithelium which has the capacity to store Vitamin A-ester (all trans) and to isomerize this compound into Vitamin A-ester (11-cis form). The retinal pigment epithelium also acts to trap incoming light as well as absorb some of the heat energy of the incoming light.

the bipolar cell, the visual information is transferred across another synaptic junction (inner plexiform layer) to the retinal ganglion cell. The axons of the retinal ganglion cells, carrying pupillomotor information (estimated at 20% of all visual fibers in the optic nerve) follow pathways similar to those for vision running in the optic nerve, optic chiasm and on to the optic tracts. The axons of ganglion cells located in the nasal half of the retina will decussate in the optic chiasm. However, the pupillary afferent fibers do not enter or synapse within the *lateral geniculate bodies*. Instead these fibers leave the optic tract via the *brachium of the superior colliculus* and enter the midbrain to synapse with the *pretectal nuclei* located near the *posterior commissure* (see Figs. 16 and 17). These light afferent fibers do not enter the superior colliculus at all. Some authors believe that visual information is carried from the retina to the visual cortex, etc., by one type of fiber only and that pupillary informa-

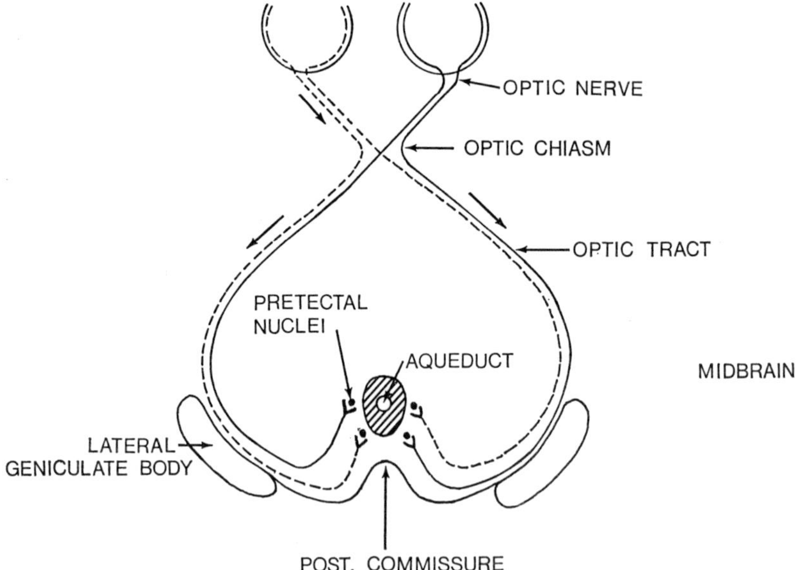

Figure 16. Afferent pathway for the light reflex. Note afferent fibers from retinal ganglion cells travel posteriorly through the optic nerve, chiasm (decussation of nasal fibers) and optic tract to synapse on the pretectal nuclei in midbrain.

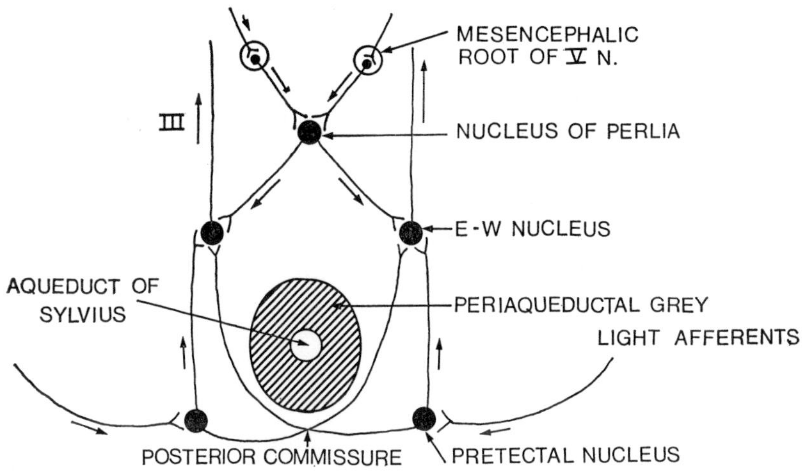

Figure 17. Parasympathetic nervous system: afferent pupillary connections within the mesencephalon, pretectal nuclei and Edinger-Westphal nuclei. (Redrawn and reproduced with permission from Duke Elder, S.: *Text-book of Ophthalmology.* London, Henry Kimpton, 1949, Vol. IV.) Note the location of the pretectal nuclei within the mesencephalon. The axons of the pretectal nuclei may synapse with the ipsilateral Edinger-Westphal nuclei or it may decussate within the posterior commisure to synapse with the contralateral Edinger-Westphal nuclei. Because of the equal decussation of fibers within the posterior commisure, the direct and consensual light reflexes are equal.

tion is transmitted to the pretectal nuclei by collaterals leaving the optic tract and entering the brachium of the superior colliculus to terminate on the pretectal nuclei.

The axons of the pretectal nuclei follow one of two pathways. First, the axons may go directly to the ipsilateral Edinger-Westphal nucleus which is located within the mesencephalon but anterior to the pretectal nucleus. Second, the axons may decussate in the posterior commissure of the midbrain and synapse with the contralateral Edinger-Westphal nucleus.

In each case, these fibers are termed *intercalated neurons* or the *pretecto-oculomotor tract.*

Occipitomesencephalic Pathway (Accommodative or Near Reflex)

Descending fibers from the occipital cortex reach the mesencephalon via a route that is very similar to that of the occipitomotor fibers going to the oculomotor complex. The near reflex fibers are believed to travel in the anterior portion of the mesencephalon in contradistinction to light reflex afferents which are running near the posterior commissure (see Figs. 17 and 26). Stimulation of these descending *occipitomesencephalic fibers* which synapse with Edinger-Westphal nuclei at the mesencephalon level results in miosis.

Inhibitory Pathways to the Edinger-Westphal Nucleus

Retinomesencephalic (Associated with the Dark Reaction)

Fibers originating in the retina travel in the optic nerve and into the optic chiasm where they leave via the *lamina terminalis* to enter the *diencephalon.* It is believed that darkness is a positive stimulus which evokes an "off reaction" in the electroretinogram which in turn gives rise to inhibitory impulses that travel from the hypothalamus to the Edinger-Westphal nucleus via *descending mesencephalic-reticular pathways,* resulting in mydriasis (see Fig. 18).

Corticothalamohypothalamic Pathway (Corticolimbic)

When a state of consciousness exists, impulses travel along the *corticothalamohypothalamic pathway* to inhibit the mesencephalic pupillary outflow emanating from the Edinger-Westphal nucleus. When a state of sleep, fatigue or unconsciousness exists, there are decreased inhibitory impulses bombarding the Edinger-Westphal nucleus from the cortex, hypothalamus and reticular activating system (see Fig. 18). This results in miotic pupils which still retain their normal reaction to light.

Reticular Activating System (Ascending Spinoreticular) Pathway

Ascending spinoreticular pathways are probably located in the anterolateral funiculus, along with the descending sympathetic fibers coming from the hypothalamus (Fig. 9), and produce a direct inhibition on the Edinger-Westphal nucleus outflow (see Fig. 18).

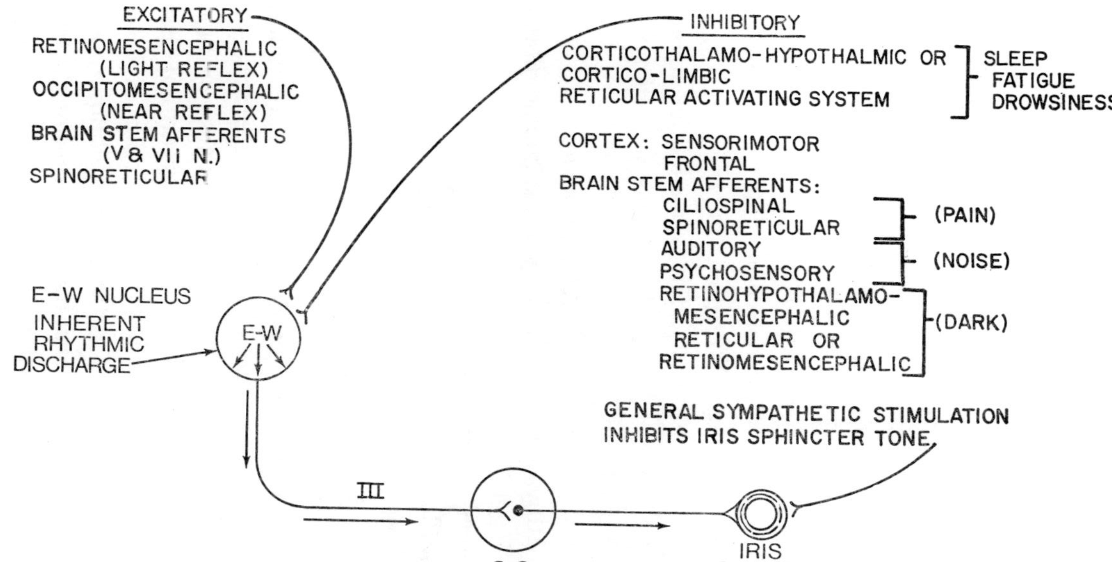

Figure 18. Excitatory and inhibitory pathways on the Edinger-Westphal nucleus. Note that the Edinger-Westphal nucleus has its own inherent constrictor tonus or rhythmic discharge that maintains the tone on the sphincter pupillae. Excitatory pathways impinging on the E-W nucleus increase miosis. Inhibitory pathways impinging upon the E-W nucleus tend to decrease the inherent constrictor tonus, resulting in mydriasis.

Efferent Parasympathetic Nervous System

Edinger-Westphal Nucleus and the Mesencephalic Outflow

The final common pathway to the sphincter pupillae consists of a two-chain neuronal circuit as follows:

The Edinger-Westphal Nucleus and Axon

The Edinger-Westphal nucleus (first neuron) is part of the oculomotor complex which is located within the mesencephalon. The nucleus has its own inherent constrictor tonus or rhythm which is spontaneous and maintains the tone on the sphincter pupillae muscle (see Fig. 18). There are numerous excitatory and inhibitory nueroanatomical pathways that impinge upon the Edinger-Westphal nucleus. It is obvious that many influences are bombarding the Edinger-Westphal nucleus at any one moment. The summation of these various modalities upon the inherent constrictor tonus of the E-W nucleus will dictate the amount of parasympathetic or mesencephalic outflow to the ciliary ganglion and hence to the sphincter pupillae.

The outflow pathway is as follows: preganglionic fibers or axons of the Edinger-Westphal nucleus exit the brain stem along with the oculomotor fibers supplying the extraocular muscles. The pupillomotor fibers are situated superficially on the dorsomedial aspects of the IIIrd cranial nerve as it exits the mesencephalon. As the oculomotor nerve proceeds towards the cavernous sinus, the pupillomotor fibers have changed their position with respect to the rest of the nerve and are found in the medial aspect near the surface. Within the cavernous sinus, the pupillomotor fibers spread out to cover the entire surface of the oculomotor nerve. Once inside the superior orbital fissure, however, the pupillomotor fibers follow the inferior division of the oculomotor nerve which goes to the inferior oblique muscle. The fibers leave that division to form the *motor root of the ciliary ganglion* (see Fig. 19). After synapsing in the ciliary ganglion, the postganglionic fibers, which are myelinated, pass to the iris and ciliary body via the short ciliary nerves (see Fig. 19). The fibers terminate on the sphincter pupillae. Fibers for accom-

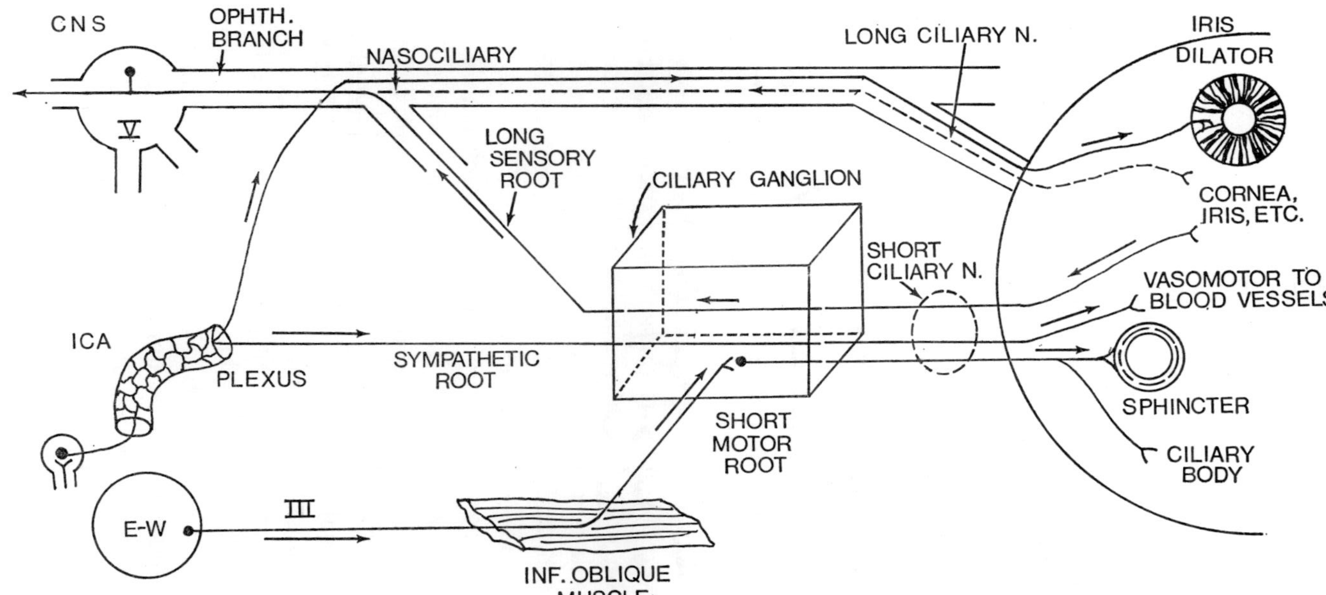

Figure 19. The ciliary ganglion. The ciliary ganglion is a parasympathetic substation where preganglionic fibers from the Edinger-Westphal nucleus synapse. The postganglionic fibers enter the globe via the short ciliary nerves to innervate the sphincter pupillae. Sympathetic fibers to the dilator bypass the ciliary ganglion, whereas sympathetic fibers with vasomotor function pass through the ganglion without synapse. (See text for further details.)

modation follow a similar pathway, synapsing in the ciliary ganglion, but terminate in the ciliary body instead.

The Ciliary Ganglion and Axon

*The ciliary ganglion (second neuron) is a parasympathetic
substation for the mesencephalic outflow of the Edinger-Westphal
nucleus.* It is approximately 1x2 mm and is embedded in loose
fatty tissue within the orbit some 10 mm anterior to the superior
orbital fissure. It is located just lateral to the ophthalmic artery,
yet medial to the medial rectus muscle. The ciliary ganglion has
three roots, the sensory, the sympathetic and the parasympathetic
motor roots, which enter posteriorly as described below (see
Fig. 19).

Sensory information from the cornea, iris and ciliary body
travel posteriorly through the short ciliary nerves, ciliary ganglion
(without synapse) and through the *long sensory root* toward the
nasociliary division of the ophthalmic branch of the trigeminal
nerve (V).

The sympathetic outflow (postganglionic) which has arborized
within the cavernous sinus is present on various blood vessels
as they enter the orbit. These sympathetic fibers leave the vessels
and enter the ciliary ganglion via the *sympathetic root* and subsequently pass through the ganglion without synapse. From the
ciliary ganglion, the sympathetic fibers enter the globe via the
short ciliary nerves and innervate the blood vessels and *branched
melanocytes* of the uveal tract. The sympathetics that go to the
blood vessels have a vasomotor function. A small number of
these sympathetic fibers go on to innervate the dilator pupillae,
even though the major pathway for dilator innervation is via the
nasociliary and long ciliary nerves.

The short parasympathetic motor root carries the pupillomotor
fibers from the Edinger-Westphal nucleus into the ciliary ganglion
where they synapse (see Fig. 19). The postganglionic fibers leave
via the short ciliary nerves to innervate the sphincter pupillae
and the ciliary body. There are from 12 to 20 short ciliary nerves
that pierce the sclera in a ring about the optic nerve. In addition,
there is a separate nerve that accompanies the central retinal

artery into the globe. Finally, studies of the ciliary ganglion reveal that 94 percent of the nuclei and emerging fibers that travel in the short ciliary nerves are concerned with accommodation, while only 3 percent deal with pupillomotor functions.

Chapter 2

PHARMACOLOGY OF THE AUTONOMIC NERVOUS SYSTEM IN RELATION TO THE PUPILLARY PATHWAYS

THIS SECTION OF the manual will deal briefly with the mechanisms of action of various autonomic drugs acting at specific sites along the pupillary pathways. Only the more commonly used clinical pharmacologic agents will be presented in detail. For a more complete listing of agents that produce either miosis or mydriasis, the reader is referred to Appendices 1 and 3.

PHARMACOLOGY OF THE SYMPATHETIC PUPILLARY PATHWAYS

The sympathetic nervous system has tracts running within the central nervous system, i.e. descending fibers from the posterior and lateral nuclei of the hypothalamus which synapse onto the intermediolateral grey column (C_8 to T_2). The neural transmitter at this synapse (see Figs. 8, 9 and 20) is *acetylcholine*. Axons, whose cell nuclei are located within the grey horn, leave the spinal cord to enter the paravertebral sympathetic chain and ascend to the superior cervical ganglion where they engage in a synaptic relationship with another neuron (see Fig. 10). This junction is termed a *ganglionic synapse*, and the neural transmitter substance is acetylcholine. *Preganglionic axons* bring neural information to the ganglion, while *postganglionic* neurons transmit this information away from the ganglion. The nerve fibers in this system have terminal endings which balloon-out and are called *boutons*. Within the *boutons*, there are *microvesicles* which contain the neurotransmitter substance, i.e. acetylcholine (see Fig. 20).

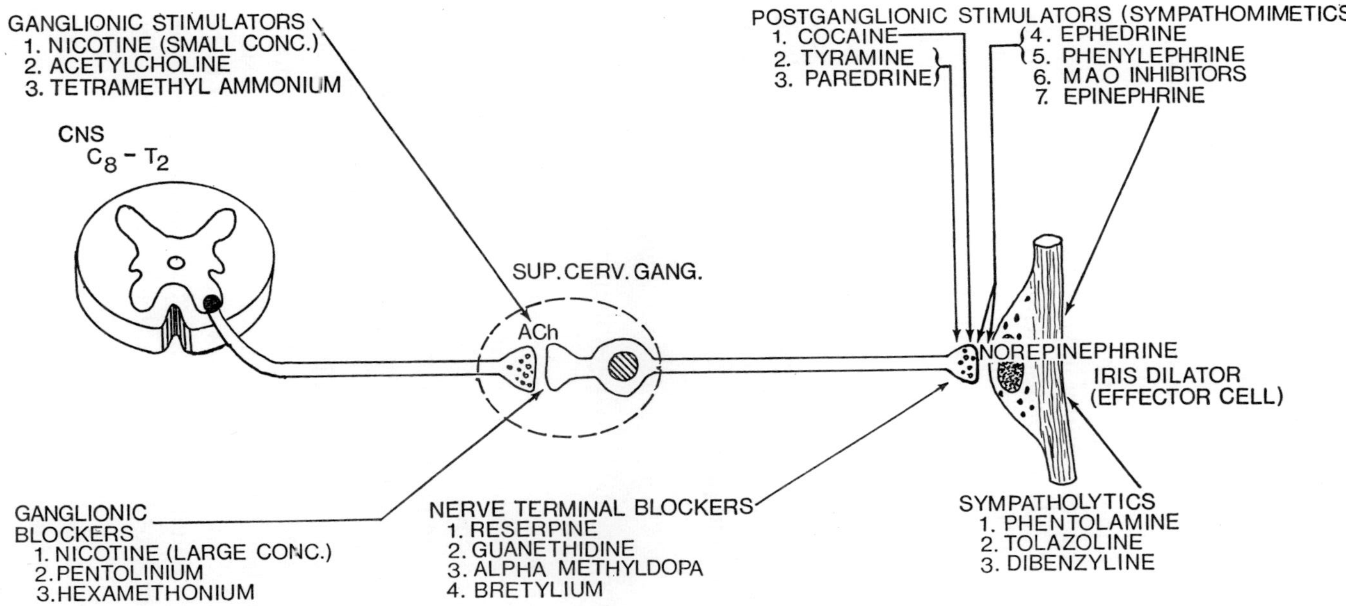

Figure 20.   Pharmacology of the sympathetic nervous system. Note that the transmitter substance for the superior cervical ganglion is acetylcholine which is metabolized by cholinesterase, whereas the transmitter substance at the postsynaptic nerve terminal is norepinephrine which is metabolized by monoamine oxidase and catechol-0-methyl transferase. (From Walsh, F. B., and Hoyt, W. F.: *Clinical Neuro-Ophthalmology*, 3rd ed. Baltimore, Williams & Wilkins, 1969.)

Postganglionic fibers leaving the superior cervical ganglion travel a long, circuitous route to reach the orbit, globe and finally the iris dilator muscles with which they synapse. However, at this synaptic junction, the neural transmitter substance is *norepinephrine,* which is stored in microvesicles within the bouton. After the release of the neural transmitter at the synaptic junction and subsequent stimulation of the postsynaptic cell, an enzymatic degradation of the transmitter takes place. Acetycholine is inactivated by *acetylcholinesterase* while norepinephrine is inactivated by *catechol-O-methyl transferase* (COMT) and/or *monoamine oxidase* (MAO). The norepinephrine that is not degraded by these enzymes is returned to and stored in the terminal bouton of the presynaptic axon.

Pharmacological agents that stimulate the superior cervical ganglion or the effector cells (dilator pupillae) will produce *mydriasis;* whereas the pharmacologic agents that block the superior cervical ganglion or the synaptic transmission at the effector cell level will produce a *miosis* (Fig. 20).

Sympathomimetic Agents

I. Superior-cervical-ganglion-stimulating agents.
   A. Acetylcholine.
   B. Nicotine (small concentrations).
   C. Tetramethyl ammonium.
II. Postganglionic- and effector-cell-stimulating agents.
   A. Cocaine (2 to 5% solution) is used as a local anesthetic as well as a mydriatic in the pharmacological evaluation of a Horner's syndrome. However, a 10% cocaine solution, when instilled into the conjunctival sac, may anesthetize the parasympathetic pupillomotor fibers to the iris sphincter, resulting in mydriasis. The mechanism of action of cocaine is believed to be one of potentiation of the transmitter substance, i.e. norepinephrine by preventing rebinding of the norepinephrine at the presynaptic nerve terminal.
   B. Tyramine and paredrine (OH-Amphetamine) release norepinephrine, when present, from the nerve endings.

C. Ephedrine and Phenylephrine (Neo-Synephrine®) not only release the norepinephrine from the nerve terminals but also stimulate the iris dilator cells directly.

D. Epinephrine (Adrenalin®) stimulates the dilator cells directly. Epinephrine in 1:1000 dilution, when applied to the conjunctiva of a normal eye, will not affect the size of the pupil. However, some authors believe that it will dilate the miotic pupil in a Horner's syndrome (see section on Horner's syndrome, pharmacological testing).

Ganglionic Blocking Agents

These agents can block the superior cervical ganglion of the sympathetic nervous system as well as the ciliary ganglion of the parasympathetic nervous system. The final size of the pupil, when these agents are administered, will depend upon the concentration and location of these agents to the various ganglia.

Superior-Cervical-Ganglion-Blocking Agents

The superior-cervical-ganglion-blocking agents are as follows:
1. Hexamethonium.
2. Nicotine (large concentrations).
3. Pentolinium.

Sympatholytic Agents

Postganglionic Blocking Agents

These pharmacological agents block the postganglionic sympathetic nerve terminal by interfering with the release of the norepinephrine. This results in depletion of the norepinephrine stores at the nerve terminal which, in turn, leads to miosis. The postganglionic blocking agents are listed as follows:

1. Guanethidine (Ismelin®).
2. Bretylium (Darenthinin®).
3. Reserpine.
4. Alpha-methyldopa (Aldomet®).

Effector-Cell-Blocking Agents

Although the agents listed below are not used frequently in clinical ophthalmology, instillation into the conjunctival sac results in miosis and ptosis. Their mechanism of action involves the blockage of the alpha receptor sites in the effector cells, i.e. iris dilator. The iris has almost exclusively alpha receptor adrenergic activity, while the ciliary body appears to have beta adrenergic receptor activity. The effector-cell-blocking agents are listed as follows:

1. Dibenzyline.®
2. Phentolamine (Regitine®).
3. Tolazoline (Priscoline®).

PHARMACOLOGY OF THE PARASYMPATHETIC PUPILLARY PATHWAYS

The parasympathetic nervous system, like the sympathetic nervous system, has tracts running within the central nervous system, i.e. descending occipitomesencephalic fibers which synapse with the Edinger-Westphal nucleus in the midbrain. The neural transmitter at this synapse is *acetylcholine.* Edinger-Westphal cell axons leave the midbrain to travel in the cavernous sinus and then enter the orbit where they enter the ciliary ganglion and synapse with another neuron. *The neural transmitter at the ciliary ganglion synapse is also acetylcholine.* Postganglionic fibers leave the ciliary ganglion (see Figs. 19 and 21), enter the globe via the short ciliary nerves, and synapse with the sphincter pupillae. The neural transmitter at the effector-cell junction is also acetylcholine. In each case the acetylcholine is enzymatically inactivated by acetylcholinesterase.

Parasympathomimetic Agents

Ciliary-Ganglion-Stimulating Agents

Ciliary-ganglion-stimulating agents are as follows:
1. Acetylcholine.
2. Nicotine (small concentrations).
3. Tetramethyl ammonium.

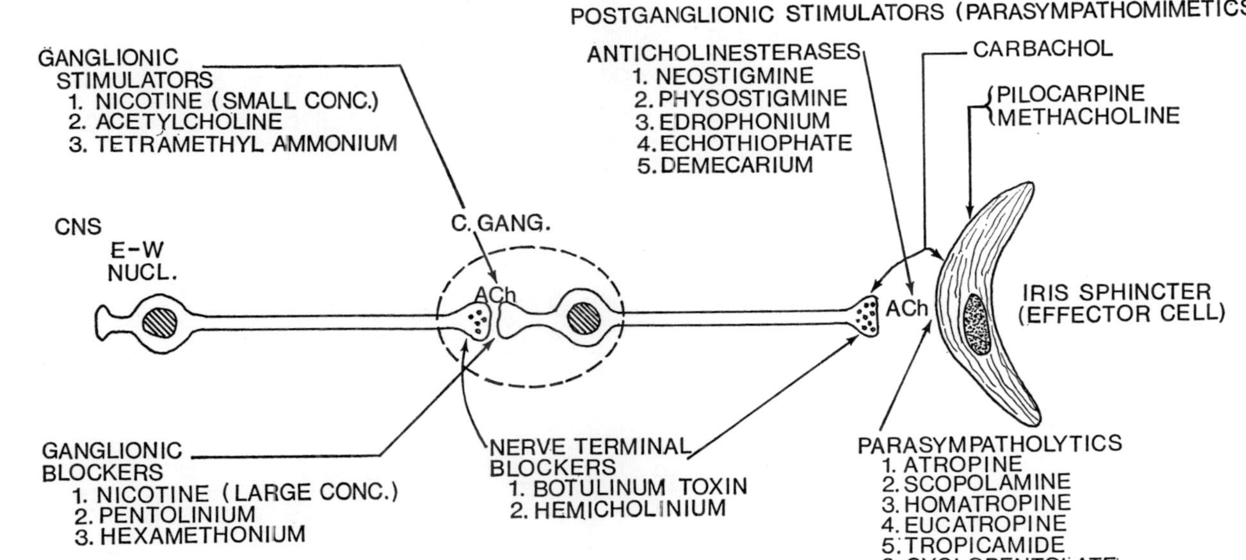

Figure 21. Pharmacology of the parasympathetic nervous system. Note that the neural transmitter substance across the ciliary ganglion synapse, as well as at the nerve terminal junction, is acetylcholine which is metabolized by cholinesterase. (From Walsh, F. B., and Hoyt, W. F.: *Clinical Neuro-Ophthalmology*, 3rd ed. Baltimore, Williams & Wilkins, 1969.)

Postganglionic and Effector-Cell-Stimulating Agents

The postganglionic and effector-cell-stimulating agents are as follows:

1. Pilocarpine, Methacholine (Mecholyl®) are true parasympathomimetic agents in that they are very similar, structurally, to acetylcholine and they stimulate the iris sphincter cells directly.
2. Carbachol (Doryl®) releases acetylcholine from the nerve terminals and also stimulates the effector cell, i.e. iris sphincter, directly.

The following agents potentiate the action of acetylcholine, when present at the nerve terminals, because they interfere with the normal degradation of the neural transmitter substance by cholinesterase. These agents are, therefore, *anticholinesterases* and can be blocked by Protopam® (pyridine-2-aldoxime methiodide). The effector-cell-potentiating agents are listed as follows:

1. Physostigmine (Eserine®).
2. Neostigmine (Prostigmin®).
3. Edrophonium (Tensilon®).
4. Echothiophate (Phospholine Iodide®).
5. Demecarium (Humorsol®).
6. Isoflurophate (DFP).
7. Tetraethylpyrophosphate (TEPP).
8. Hexaethyltetrophosphate (HETP).
9. Octamethylpyrophosphoramide (OMPA).

The continued use of these parasympathomimetic agents, i.e. pilocarpine, may lead to a condition known as rigidity of the iris, which is a state of fixed contraction of the sphincter pupillae. The reactions of the pupil, namely mydriasis and miosis, may be impaired. If this condition persists, posteror synechiae may form.

Another untoward effect of long-term miotics, especially the use of DFP or Phospholine Iodide, is the *formation of cysts at the pupillary margin*. This is often seen in children on these agents for strabismus problems, i.e. accommodative squint. The cysts are believed to be caused by the contraction of the sphincter against the lens capsule, thereby pinching the iris pigment epithelium. With continued use of the miotics, the cysts will grow and possibly rupture. When miotics are discontinued, the cysts gradually disappear. Neo-Synephrine (10% or 2.5% solution) has been known to diminish the size of these cysts.

When a patient is on anticholinesterase medications and he is scheduled for surgery with general anesthesia and muscle relaxation is important, *it is imperative that succinylcholine not be given, for prolonged apnea and death may result.* Curare may be used in these cases for muscle relaxation. A period varying from three weeks or more after discontinuing the anticholinesterase may be necessary before succinylcholine may be safely administered.

Ganglionic Blocking Agents

See "Ganglionic Blocking Agents" in the drug section on the "Sympathetic Nervous System."

The following are ganglionic blocking agents:

1. Hexamethonium.
2. Nicotine (large concentrations).
3. Pentolinium.

Parasympatholytic Agents

Postganglionic Blocking Agents

1. Botulinum toxin blocks the release of acetylcholine at the ganglion and effector-cell levels.
2. Hemicholinium interferes with the synthesis and release of acetylcholine at the ganglion and effector-cell levels.

These two pharmacologic agents also block nerve transmission in the ganglia of the sympathetic nervous system where acetylcholine is the transmitter substance.

Effector-Cell-Blocking Agents

These agents are similar in their molecular configuration to acetylcholine in at least that portion of the molecule that is involved in receptor site stimulation, so that these agents compete with acetylcholine for the receptor site on the iris sphincter muscle. Because these agents do not stimulate the iris sphincter cell when they attach to the receptor site, they, in effect, block the parasympathetic outflow at the effector-cell level. The effector-cell-blocking agents are listed as follows:

1. Atropine.
2. Scopolamine.
3. Homatropine.
4. Eucatropine.
5. Tropicamide (Mydriacil®).
6. Cyclopentolate (Cyclogyl®).

Miscellaneous Pharmacologic Agents

1. Histamine is an irritant and can affect the iris sphincter directly producing a miosis. This effect will occur even if the sphincter is fully atropinized!
2. Morphine blocks descending inhibitory cortical fibers to the Edinger-Westphal nucleus (see Fig. 18) which results in profound miosis. Some authors believe that it may also have a direct stimulating effect on the Edinger-Westphal nuclei. However, topical morphine has little effect on the pupil.
3. Nalorphine (Nalline®) is an antinarcotic agent and when given subcutaneously can reverse the miotic action of morphine!
4. Steroids when topically administerd, i.e. dexamethasone, have a mild mydriatic effect. The mechanism for this is not clear.
5. LSD-25 will evoke mydriasis in the prodromal stage.

Serum Cations

1. Low serum calcium. Calcium ions tend to facilitate the release of acetylcholine. When there is a low serum calcium, synaptic transmission across junctions normally mediated by acetylcholine become blocked. This may result in mydriasis.

2. High serum magnesium tends to block acetylcholine mediated synaptic junctions. This may result in mydriasis.

A further listing of toxic agents that produce mydriasis or miosis may be found in Appendix 2.

Chapter 3

FACTORS THAT INFLUENCE
PUPILLARY SIZE

IN GENERAL THE pupil varies in size from 2 to 5 mm in diameter with extremes of 0.5 mm to 9.0 mm. Because of the magnifying power of the cornea the pupil appears one eighth larger than it really is. In the general population, the majority of patients have pupils that are equal and symmetrical. However, about 17 percent of normal patients have a slight inequality of their pupils which is termed *anisocoria*. About 4 percent of the population has pronounced anisocoria (greater than 1 mm difference between the diameters of the pupils). The latter individuals should be examined thoroughly, including pupillography, to rule out disease. The factors that influence the size of the pupil are mechanical, extrinsic or intrinsic.

I. Mechanical factors.
   A. A brown iris which is rich in pigment will usually have a smaller pupil than a blue iris (poor in pigment). It is also a well-known clinical fact that mydriatic agents are less effective in dilating darkly pigmented irides compared to blue irides.
   B. Myopes have larger pupils than hyperopes. Hyperopes have larger ciliary bodies than myopes and so present a shorter diameter for the iris root insertion, which in turn results in a smaller-sized pupil.
   C. Local anatomic lesions of the iris musculature, i.e. rupture or paresis of the iris sphincter, synechiae, sclerosis of iris sphincter in old age, and iridodialysis, etc., affect the size of the pupil. The neuroanatomical lesions along various points of the pupillary pathways and their effects on the pupils will be discussed in Chapter 6.

II. Extrinsic factors.

    A. The rate of change in illumination per unit of time that is presented to the retinal receptors is of primary importance in the initiation of the pupillary reflex to light. If the light presented to the retinal receptors remains constant in intensity, then miotic pupils will enlarge somewhat and reach a new equilibrium point. The more intense the light stimulus, the greater will be the resultant miosis.

    B. For a given light stimulus, a dark-adapted eye will undergo a more vigorous miosis than a light-adapted eye.

    C. The distance of the object of fixation will, in part, determine the pupillary size, i.e. a near object stimulates convergence as well as miosis. However, if the object is closer to the eye than the nearpoint of convergence for that individual, then one eye will deviate outward with a relaxation of convergence, and the pupils begin to dilate.

III. Intrinsic factors.

    A. Increase in sympathetic tonus results in mydriasis, whereas a decrease in the sympathetic tonus results in miosis. In addition, the increased sympathetic tonus or outflow will inhibit iris sphincter tone thereby facilitating mydriasis. This inhibition is believed to take place at the iris sphincter directly (see Fig. 18).

    B. The *Edinger-Westphal nucleus* has its own inherent rhythmic discharge that maintains a baseline tonus on the iris sphincter. Superimposed upon this constrictor tonus are various excitatory and inhibitory impulses (see Fig. 18).

    C. In the newborn or infant, the pupils are miotic and remain that way through the first year of life. During childhood and adolescence, the pupils are about maximum size, and with advancing age they gradually become smaller. In elderly patients, the pupils are often miotic, as they are in infancy.

D. Intense emotion—joy as well as horror—can produce mydriasis.
E. Psychoneurotic depression often results in miosis.
F. Sleep is consistent with low sympathetic tonus as well as decreased inhibition of the Edinger-Westphal nucleus resulting in miosis.
G. General anesthesia.

   1. Stage I is an irritative stage with emotional stress and a resultant mydriasis.
   2. Stage II is also irritative with resultant mydriasis.
   3. Stage III.
      a. In planes 1 and 2, the pupils are contracting and are about the size of pupils in deep sleep, or miotic.
      b. In Planes 3 and 4, the pupils undergo mydriasis. This may be a reflection of generalized depression of spinal cord and mesencephalon structures, i.e. Edinger-Westphal nucleus.
   4. In stage IV, the pupils suddenly dilate widely and often are unresponsive to strong light stimuli. Again this mydriasis may be due to obtundation of mesencephalic structures including the Edinger-Westphal nucleus. Of course the diencephalic structures are also depressed.

H. Respiratory movements. It has been shown that the rate and quality of pupillary oscillations have no direct relationship to the various phases of the normal respiration, i.e. inspiration and expiration.
I. Pupillary unrest. The pupils of patients whose eyes are normal may be seen to be in constant motion, using ordinary illumination. When light is presented to a normal eye, there is an initial miosis with a momentary stabilization of pupillary size followed by a partial redilation and then small oscillations of the pupil. As the intensity of the light stimulus increases, the magnitude of these oscillations also increases. This *physiologic pupillary unrest* is found in normal individuals, but is

reduced or absent in lesions of the afferent or efferent limb of the light reflex (see Chapter 6) or in the presence of spastic miosis. Pupillary unrest is usually symmetrical. The term *hippus* is reserved for conditions when there is a high degree of physiological pupillary unrest. True hippus is a rare clinical entity and has no localizing significance (see Appendix 1). It has been observed under the following clinical conditions: multiple sclerosis, epilepsy, meningitis (acute), cerebral tumors, incipient cataracts, barbiturate and paraldehyde poisoning and normals. Hippus may be due to spontaneous fluctuation of the accommodative reflex.

The size of the pupil as well as the excursion of the iris or pupillary movements are determined by numerous conditions and influences as outlined above. Because the sphincter muscle is by far the stronger of the two intrinsic muscles of the iris the variability in pupillary size is mainly due to the forces that influence the Edinger-Westphal constrictor tonus.

Chapter 4

NORMAL PUPILLARY REFLEXES

PUPILLARY REFLEXES THAT PRODUCE MIOSIS

Light Reflex

THERE ARE NUMEROUS factors involved in the light reflex which begins with the stimulation of the retinal receptors, i.e. rods and cones, by incoming light. These factors are partially listed below.

1. The primary factor which governs whether an incoming light stimulus will evoke pupillary miosis is the background adaptation state of the retina, for it is the rate of change in illumination per unit of time that appears to be the key to triggering this reflex. If the illumination change is very slow and stretches over a long time interval, then the amount of pupillary reaction might be negligible. However, if there is an abrupt rise in the illumination presented to the retina, then one would expect a brisk miosis to ensue.

2. The area of the retina that is stimulated by the incoming stimulus is also important. The macular and foveal regions are much more sensitive to changes in illumination and therefore will respond with strong pupillomotor outflow on minimal stimulation. On the other hand, the peripheral retina as well as the retina immediately adjacent to the optic nerve head are much less sensitive to changes in illumination.

3. The direction of incoming quanta of light energy may have a theoretical influence upon the light reflex. The cone receptors are highly sensitive to the direction of incoming light and will fire only when the light is directed down the main or longitudinal axis of the cell. Oblique pencils of light that strike the cone receptor will not stimulate the cell to fire. Rod receptors, on the other hand, will fire when parallel rays (to the long axis

of the receptor cell) as well as oblique rays strike the receptor portion. There are 120 million or more rods per retina, as opposed to 7 million cones per retina. Actually, incoming light undergoes internal scatter, so that there is often diffuse stimulation of the retina, and pupillary response is a combined rod-and-cone affair. However, if one could direct a small pencil of light onto the optic disk and thereby avoid hitting any of the cone cells in a line parallel to their longitudinal axis, the pupillary response would be entirely due to the rod receptors!

4. Level of consciousness, fatigue and emotional excitement—are important modifying influences on the light reflex. In cases of sleep and fatigue, there is a decrease in the descending inhibitory cortical discharges affecting the Edinger-Westphal nucleus (see Fig. 18). Under these conditions, the pupils tend to be miotic; when a bright light stimulus is presented, there is not much of a pupillomotor response. In states of emotional excitement, on the other hand, the pupils are large, reflecting increased cortical inhibitory influences impinging upon the Edinger-Westphal nucleus. Under these conditions, a strong light stimulus will evoke a large miotic reaction.

5. There is a *spectral sensitivity* of the retina to incoming light. For example, in the dark-adapted eye, the maximum pupillary response is found with light of 500 mμ, whereas in the light-adapted eye the maximum response is found with light of 560 mμ. This spectral sensitivity of the pupil has been termed the *Pürkinje shift* (see Fig. 22). Of course this assumes that equal amounts of light stimuli are presented to the eye.

Neuroanatomical Pathway for the Light Reflex

Direct Light Reflex

Rod or cone receptor cells fire on appropriate light stimulation and relay the neural information to the ganglion cells via bipolar cells or other interconnecting cells (see Fig. 15). From the ganglion cell, the messages are transmitted along the optic nerve, with 50 percent of the fibers decussating (those originating in the nasal half of each retina) at the optic chiasm. The information then continues posteriorly in the optic tracts to the pretectal

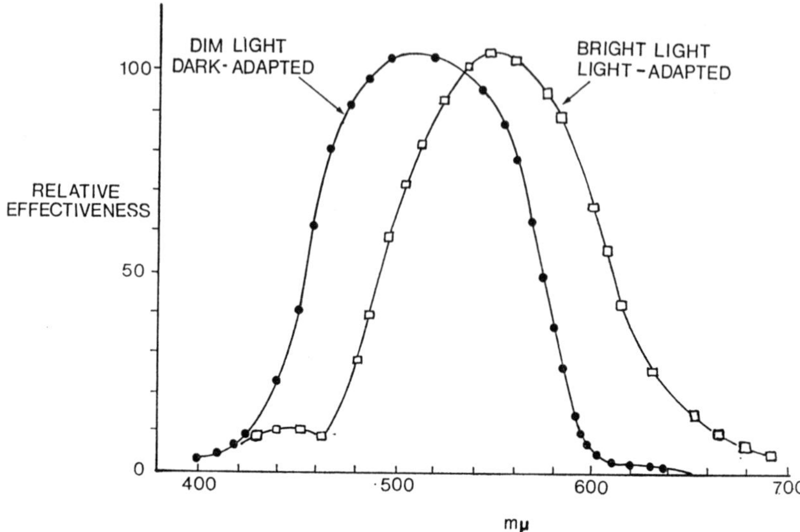

Figure 22. Pupillary spectral sensitivity: Pürkinje shift. Note that the vertical axis in the graph represents relative effectiveness of pupillary response. The horizontal axis represents the different wavelengths of light stimuli presented to the retinal receptors. (From Laurens, H.: Studies on the relative physiological values of spectral lights. III. The pupillomotor effects of wave-lengths of equal energy content. *Am. J. Physiol.*, vol. 64, No. 1, March 1, 1923.)

nuclei in the mesencephalon (see Figs. 16 and 17). Here intercalated neurons synapse with the Edinger-Westphal nuclei. Some of the pretectal cells send axons to ipsilateral E-W nuclei while others are found decussating in the posterior commissure to synapse with contralateral E-W nuclei. The efferent arm of this reflex commences with the Edinger-Westphal nuclei which sends out axons in the IIIrd nerve complex. The efferent pupillary fibers synapse in the ciliary ganglion (ipsilateral), and post-ganglionic fibers (short posterior ciliary nerves) enter the globe and run in the suprachoroidal space, finally synapsing with the iris sphincter muscle, stimulation of which results in miosis.

The latent period for this reflex varies from 0.18 to 0.20 sec. The time at which maximal miosis is observed is about 0.94 sec. after the presentation of the light stimulus.

Consensual Light Reflex

The afferent loop of this pathway is the same as the direct reflex. The only difference is that the decussation in the posterior commissure to the contralateral Edinger-Westphal nucleus is involved, so that miosis which is equal in intensity, timing and quality to that of the direct reflex (in the eye directly stimulated) is seen in the nonstimulated eye. This reaction is termed the *consensual light reflex.*

Accommodation-Convergence Reaction (Near-Reflex or Near-Point Fixation)

The sphincter pupillae, ciliary body and medial rectus muscle are all innervated by the third cranial nerve. When a person looks at an object which is relatively near to him, the following series of synchronous events takes place:

1. The medial recti are activated to adduct the eyes so that the image of the object will be on each fovea simultaneously.
2. The ciliary body contracts, allowing increased curvature of the lens to take place, bringing the image rays onto the fovea.
3. The pupillary sphincter muscle is also activated and miosis ensues with a resultant increase in the depth of focus. This is termed the near-point reaction of the pupil.

Each of the preceeding reactions is independent of the other two, and therefore the entire coordinated series of events cannot be truly labeled a reflex but rather a *synkinesis.* In fact, some investigators have shown that the miosis is not related to the accommodation effort at all but to the convergence mechanism. In support of this idea is the fact that miosis is observed in myopic patients and in the elderly who have lost their accommodation. In addition, it is known that pupillary miosis begins when an approaching object is 0.5 meters away from the patient and not when his accommodation efforts commence. Finally, it is a well established clinical observation that as an object is brought closer to the subject than his *near point of convergence* his eyes will diverge and pupillary mydriasis will take place (see p. 45).

Afferent Pathway

Rod and cone cells, sending information to other intraretinal cells, eventually stimulate the retinal ganglion cells. Visual information is relayed to the occipital cortex via the optic nerve, optic chiasm, optic tract, lateral geniculate body and optic radiations, in that order. After complex interneuronal association relays, there are descending tracts which make their way to the anterior portion of the mesencephalon and synapse with the Edinger-Westphal and oculomotor nuclei complex.

Efferent Pathway

From the Edinger-Westphal nucleus, pupillomotor fibers run forward (preganglionic), passing through the cavernous sinus, superior orbital fissure, and then the *annulus of Zinn* to enter the orbit as part of the *inferior division of the oculomotor nerve,* finally synapsing in the ciliary ganglion (see Fig. 19). Warwick states that 94 percent of the synapsing fibers within the ciliary ganglion are related to accommodative function, whereas only 3 percent are involved in pupillomotor functions. The postganglionic pupillomotor fibers innervate the sphincter pupillae and the accommodative fibers innervate the ciliary body. The medial recti are innervated by other oculomotor fibers.

Lid-Closure (Orbicularis) Reflex

When a subject closes his eyelids, the pupils become miotic. This can be observed clinically by asking the patient to close his eyes while the eyelids are held open. The miosis that occurs is termed the *orbicularis reflex* and this presumes a normal innervation of the orbicularis muscle. In addition, the globe rotates upward (*Bell's phenomenon*) as part of the lid-closure reaction.

In cases where the pupil fails to react to light and accommodative effort, the orbicularis reflex is helpful in establishing the integrity of the efferent pupillary pathway, i.e. a positive lid-closure reaction.

Afferent Pathway

The afferent pathway is not known.

Efferent Pathway

The efferent pathway is the same as that described in the preceeding section for the near reaction.

The existence and nature of the orbicularis reflex has been used by several authors as partial proof for the existence of direct connections between the seventh nerve nucleus and the oculomotor complex, including the Edinger-Westphal nucleus.

The lid-closure reflex takes place whether there is a voluntary or a reflex closure of the eyelids as occurs in blinking and in corneal irritation. When the reflex is present (not all normals have this reflex miosis), it occurs ipsilaterally. No consensual miosis takes place.

Trigeminal (Oculopupillary) Reflex

It is a well-established clinical observation that any irritation of the cornea, conjunctiva or eyelids results in constriction of the pupil. Painful stimuli lasting a brief time will result in mydriasis. If the pain persists, there will be a resultant miosis, which is bilateral. However, the pupil on the affected side will be more miotic.

Afferent Pathway

The afferent pathway consists of stimulation of the ophthalmic branch of the fifth cranial nerve which sends information via the gasserian ganglion and then on to the nucleus of the trigeminal nerve. Then internuncial fibers go via the *medial longitudinal fasciculus* to the Edinger-Westphal nucleus.

Efferent Pathway

The efferent pathway is from the Edinger-Westphal nucleus to the ciliary ganglion with a synapse taking place. Postganglionic fibers terminate in the iris sphincter (see Fig. 19).

Axon Reflex

A second mechanism or pathway for miosis secondary to painful stimuli is noted below.

Afferent Pathway

The afferent pathway is trigeminal nerve stimulation with impulses going toward the spinal cord from the periphery, i.e. cornea. However, there is an antidromic flow of these impulses to blood vessels of the iris, resulting in vasodilation of the iris capillaries. This is termed the *axon reflex*. Dilation of iris blood vessels results in miosis.

Efferent Pathway

Once the afferent impulses begin to travel in an antidromic fashion, the "efferent arc" of this reflex is established. There is no synapse between the afferent and so-called "efferent" limbs of this reflex.

It is also noted that in patients with *iridocyclitis*, there is a miotic pupil which, in many cases, does not respond to atropine very well. As the inflammation subsides, the iris tissue responds to the atropine, and mydriasis ensues.

PUPILLARY REFLEXES THAT PRODUCE MYDRIASIS

Dark Reflex

Withdrawal of light from the environment causes a reflex dilatation or mydriasis which is based on two mechanisms. First, there is a passive relaxation of the iris sphincter secondary to inhibitory impulses impinging upon the Edinger-Westphal nucleus. Second, there is an active contraction of the dilator pupillae via the cervical sympathetic pathway.

Psychosensory Reflexes

These reflexes require sensory stimulation (touch, pain, sound, etc.) which in turn is relayed to the sympathetic nervous system centers centrally, i.e. ciliospinal center of Budge, then peripherally to the superior cervical ganglion. Postganglionic fibers carry this information to the iris dilator muscle with resultant mydriasis. The various psychosensory reflexes are as follows:

1. Ciliospinal reflex works with touch stimuli applied to the neck region. However, pain stimuli evoke a better response, which is ipsilateral mydriasis. Actually pain stimuli administered at almost any site on the skin of the body can elicit this reflex.
2. Cochleopupillary reflex takes place when a vibrating tuning fork is brought close to one ear, with resultant ipsilateral mydriasis.
3. Vestibulopupillary reflex is elicited when the labyrinth is stimulated, resulting in ipsilateral mydriasis.
4. Flateau's neck mydriasis occurs when the neck is flexed.
5. Redlich's phenomenon is mydriasis secondary to any vigorous sustained muscular activity. The probable mechanism in this reflex is stimulation of the sympathetic nervous system.
6. Myer's iliac phenomenon is pressure on McBurney's point in the right lower quadrant of the abdomen, resulting in mydriasis.

Chapter 5

CLINICAL EVALUATION OF THE PUPIL

General Medical History

IN A CLINICAL evaluation of the pupil, a general medical history is obtained initially, with special emphasis on previous systemic infections, i.e. tuberculosis, syphilis, toxoplasmosis, etc., as well as prevous injuries to the head and neck region, including loss of consciousness. Queries about metabolic aberrations, i.e. diabetes mellitus, hypertension, gout, etc., should be taken. A history of drug ingestion, collyria, allergies, and smoking and drinking habits is important. A familial history of disease, with emphasis on infections, congenital defects, cancer, aberrations of the central nervous system, etc., should be recorded.

For a more comprehensive list of general medical history questions or topics, refer to any of the standard medical textbooks on the subject.

Specific Ocular History

A specific ocular history includes the following:

I. Present eye history.
    A. Pain in and about the eyes, orbits, head and neck regions.
    B. Irritation or inflammation of the eyelids, conjunctiva or areas adjacent to the eye.
    C. Secretions (tearing or discharge).
    D. Lumps, masses or swelling about the ocular region.
    E. Disturbances in the position (exophthalmos, esotropia, exotropia, hypertropia, etc.) and movement of the eyeball (paresis of extraocular muscles).
    F. Disturbances or defects of vision.
        1. Loss of vision.

    2. Decreased vision for near (hyperopia), far (myopia), central (scotomas) or peripheral.
    3. Vertigo.
    4. Diplopia.
    5. Chromatopsia.
    6. Nyctalopia (day vision or night blindness).
    7. Hemeralopia (night vision or day blindness).
    8. Photophobia.
    9. Metamorphopsia, macropsia, micropsia.
   10. Photopsia.
   11. Hallucinations.
   12. Muscae volitantes.
   13. Scintillations.

Note that in loss of vision inquiries should be made as to whether it is sudden (amaurosis fugax, temporal arteritis, etc.), gradual or intermediate. Is there a sudden loss of vision for far or distant objects (sudden myopic vision often seen in diabetes mellitus)? Is there a field loss? Is there a loss of color vision, and if so, for which colors? Is there a specific loss of ability to read (dyslexia)?

II. Past eye history.
    A. Features as noted under "present eye history."
    B. Optical history.
    C. Ocular surgery.
    D. Previous eye medications.

General Physical Examination and Neurological Examination

The details for these examinations are covered in any of the standard textbooks on physical diagnosis and will not be discussed in this manual.

Specific Ocular Examination

A specific ocular examination includes tests for:
  I. Visual acuity (with and without the pinhole) for distance vision, using best corrected vision.
 II. Visual acuity for near using best corrected vision.

III. Inspection of eyes with diffuse and focal illumination with and without a 10X loupe. Using these techniques, the following ocular structures are studied:
   A. Eyebrows, orbital rims and eyelids.
   B. Cilia, meibomian glands and semilunar folds (including caruncle).
   C. Conjunctiva, bulbar and tarsal.
   D. Puncta and lacrimal apparatus.
   E. Globes for proptosis, hypertelorism, telecanthus and microphthalmia.
   F. Cornea, sclera, limbus and anterior chamber.
   G. Iris, pupil and crystalline lens. Note the color of the irides as well as contour and symmetry of the pupils. Is there any anisocoria?
IV. Palpation of the following structures:
   A. Head, including cranial vault and temporal arteries.
   B. Preauricular, submandibular and submaxillary lymph nodes.
   C. Margins of the orbit for tenderness and the globe for resilience.
   D. Eyelids and lacrimal apparatus.
   E. Lacrimal gland.
   F. Conjunctiva.
   G. Globe. (If proptosis is present, palpate the thyroid gland.)
   H. Muscle tenderness.
V. Auscultation in various positions (sitting, lying and bending forward) for bruits in the following regions:
   A. Temporal artery region.
   B. Eyebrow region.
   C. Over the globe.
   D. Neck, for carotid artery or thyroid problems.
   E. Heart (correlation of a specific bruit with systole or diastole or identification of a cardiac bruit that is transmitted to the neck region).
   F. Antecubital fossa (blood pressure in both arms).

VI. Motor-function studies. The extraocular muscles are examined for:

    A. Primary position of gaze.

    B. Cardinal fields of gaze.

    C. Cover and uncover tests.

    D. Convergence (near point of convergence, NPC).

    E. Ocular dysmetria.

    F. Forced eye closure (Bell's phenomenon).

    G. Doll's head movements and caloric eye responses.

    H. Nystagmus (if present, see *special tests* below).

    1. Diplopia (if present, see *special tests* below).

VII. Tests for lid function are performed, looking for:

    A. Lid lag.

    B. Ptosis.

    C. Lid retraction.

    D. Blepharospasm.

VIII. Tests for pupillary function should include

    A. Light reaction

        1) Direct light reflex.

        2) Consensual light reaction.

    B. Near reflex (accommodative effort).

    C. Ciliospinal reflex.

Note that if the pupillary responses are abnormal or equivocal and the anterior-segment examination does not explain the reactions, i.e. anterior or posterior synechiae, glaucoma, etc., then refer to the section on special tests below for further evaluation procedures.

IX. Slit-lamp examination of all the anterior segment structures including the posterior lens capsule and anterior face of the vitreous body. Also note the structure and color of the iris. Look for signs of iris atrophy by transillumination. Is there any iridodonesis, ectropion uvea, or anterior or posterior synechiae formation? Are there any colobomas present? Also note the size, shape and number of the pupils in one eye and compare it to the other eye. Note the reaction of the iris or pupil to the beam of light

from the slit lamp as it is swung in and out of the pupillary
aperture.

X. Visual fields using the tangent screen or the Goldmann
perimeter.

XI. Tonometry with gonioscopy if necessary.

XII. Fundus examination (emphasis on the optic disc and
macular zones).

A. Direct ophthalmoscopy.

B. Indirect ophthalmoscopy.
1. Monocular.
2. Binocular.

C. Slit Lamp.
1. Goldmann lens.
2. Hruby lens.

XIII. Special tests, including

A. Tests for diplopia, including the red glass and Hess
diplopia screen.

B. Tests for nystagmus, including opticokinetic nystagmus (O.K.N.) drum, labyrinth stimulation, etc.

C. Exophthalmometry.

D. Ophthalmodynamometry.

E. Fluorescein angiography.

F. Skull x-rays and orbital views, including optic
foramina.

G. Complete blood count (cbc), urinalysis, VDRL, and
FTA-ABS if necessary.

H. Anterior chamber tap with dark-field examination for
spirochetes.

I. Ishihara charts for color blindness determination.

J. Carotid angiograms, electroencephalogram, brain scan,
lumbar puncture, pneumoencephalograms, etc. These
tests should be done in conjunction with members of
either the neurology or neurosurgery group in the
hospital.

K. Pupillary function tests.
1. Dark adaptation.
2. Cocaine test.

    3. Mecholyl test.
    4. Epinephrine test.
    5. Ciliospinal reflex.
    6. Orbicularis reflex.
    7. Swinging flashlight test.
    8. Pupillography.
  L. Tensilon test.
  M. Electroretinograms and evoked occipital potentials.
  N. Corneal reflexes (direct and consensual).
  O. Tests for olfaction.

Swinging Flashlight Test

In the swinging flashlight test, a flashlight is alternately swung from one eye to the other, usually in a darkened room, and if the pupillary pathways are intact, a bilateral and equally symmetrical miosis will be observed. The reason for this observation lies in the fact that there is a 50 percent decussation of the afferent pupillary fibers at the optic chiasm as well as the fact that there is an equal (50%) decussation of pretectal nuclear fibers in the posterior commissure of the mesencephalon (see Figs. 16 and 17). Therefore, clinically, the direct and consensual light reactions are equal in intensity, quality and timing.

Rapidly moving a light from one eye to another instantly substitutes a direct-light response for a consensual-light response. If there is unilateral optic nerve damage, the direct light response in this pupil is less than the consensual response obtained from the normal eye. Therefore, when a flashlight is moved from a normal eye to one with optic nerve damage, the pupil dilates and this is termed a positive swinging flashlight test or a *Marcus Gunn sign* and represents abnormal pupillary escape.

A positive swinging flashlight test is found in the following:

1. Intraocular or retrobulbar optic neuritis.
2. Optic nerve atrophy.
3. Retinal detachment.
4. Central retinal artery occlusion.
5. Central retinal vein occlusion.

6. Direct pressure on any portion of the optic nerve by either an intraorbital or intracranial mass lesion.
7. Other widespread disease of the retina.

A negative swinging flashlight test is found in the following:
1. Amblyopia exanopsia associated with squint.
2. Hysterical amblyopia (functional).
3. Papilledema secondary to increased intracranial pressure.
4. Refractive errors.

Pupillography

Pupillography can be defined as the scientific study and measurement of the pupillary response to various physiological stimuli under a variety of conditions in the normal and pathologic states. The pupillograph is a complex instrument which uses an infrared spot scanner to determine the size of the pupil at any one moment. This is possible because the heat given off by the iris can be detected by this scanner, whereas little heat is given off in the pupillary zone. In addition, this machine can integrate the total area of the pupil at any one moment in time. We owe the bulk of our knowledge about the pupillographic reactions of the pupil to Lowenstein and Lowenfeld; refer to their work for further information on this fascinating discipline.

Chapter 6

CLINICAL NEUROPATHOLOGY INVOLVING THE PUPILLARY PATHWAYS

THE CONSCIOUS PATIENT: PARASYMPATHETIC NERVOUS SYSTEM

Lesions of the Afferent Arc

THE AFFERENT ARC of the light reflex can be divided into three components, as follows:

1. Retinal chain (photoreceptor cells, i.e. rods and cones, as well as bipolar cells etc.).
2. Retinal ganglion cell.
3. Pretectal neuron.

Retinal Chain Lesions

Retinal lesions can affect the pupillary reflexes. However the pupillary changes are usually of secondary importance in localization of the problem. Ophthalmoscopic examination is the key to the clinical diagnosis. The reader is referred to the standard texts on diseases of the fundus for further details of the differential diagnosis of retinal chain lesions.

Retinal Ganglion-Cell Lesions (Optic Nerve)

The axons of the retinal ganglion cells come together in a trunk as they exit through the *posterior scleral foramen* and are collectively termed the *optic nerve* (anatomically it is a tract of the central nervous system rather than a cranial nerve). The optic nerve can be subdivided into four zones as follows:

1. Intraocular or ocular portion, which exits through the posterior scleral foramen and is 1 mm long.

2. Orbital portion (25 to 40 mm long), that follows a sinuous route to the apex of the orbit.
3. Intracanalicular portion (4 to 10 mm long), that traverses the optic canal or foramen located within the lesser wing of the sphenoid bone.
4. Intracranial portion (10 mm long) that extends from the posterior portion of the optic foramen to the optic chiasm.

In general, the optic nerve, posterior to the *lamina cribrosa*, is medullated and is about 3 to 4 mm in diameter as far as the optic foramen. The intracranial portion of the optic nerve is 4 to 7 mm in diameter.

There are numerous pathological states that can result in damage to the optic nerve, manifested as either papilledema or optic-nerve atrophy. A brief list of the etiologic agents with specific disease states is given below.

DIFFERENTIAL DIAGNOSIS OF OPTIC-NERVE DAMAGE

(PAPILLEDEMA OR ATROPHY)

The following are possible causes of optic-nerve damage.
I. Congenital.
   A. Drusen.
   B. Optic pits.
   C. Colobomas.
   D. Staphylomas.
   E. Aplasia of optic nerve.
II. Hereditary.
   A. Leber's optic atrophy.
   B. Behr's optic atrophy.
   C. Congenital or infantile hereditary optic atrophy.
III. Inflammatory (optic neuritis).
   A. Local conditions such as sinusitis, meningitis, orbital inflammation, retinitis, choroiditis and parasitic cysts.
   B. Systemic viral disease (poliomyelitis, measles, mumps, etc.).
   C. Systemic bacterial disease (septicemia, pneumonia, etc.).

IV.   Demyelinating disorders.
    A. Multiple (disseminated) sclerosis.
    B. Diffuse scleroses such as Schilder's disease, neuro-myelitis optica (Devic's disease) and leukodystrophies.
V.   Noninflammatory swelling.
    A. Pseudopapilledema (normals, hyperopia, emphysema).
VI.   Neoplasms.
    A. Fourth ventricle and cerebellar tumors (children).
    B. Meningiomas of the middle or anterior cranial fossa (adults).
    C. Pinealomas (Parinaud's syndrome).
    D. Craniopharyngiomas.
    E. Metastatic neoplasms (including paranasal sinus foci).
    F. Pituitary adenomas.
    G. Optic nerve gliomas.
    H. Chordomas.
    I. Teratomas.
    J. Lipomas.
    K. Retinoblastomas.
VII. Vascular.
    A. Aneurysms of internal carotid artery, anterior communicating artery (saccular), etc.
    B. Subarachnoid hemorrhage secondary to a ruptured aneurysm.
    C. Subdural hematoma.
    D. Vascular accidents (thromboses, emboli with infarction, etc.), temporal arteritis.
    E. Blood dyscrasias.
VIII. Systemic diseases associated with papilledema.
    A. Acute idiopathic polyneuritis (Guillain-Barré syndrome).
    B. Infectious mononucleosis.
IX.   Primary ocular disease (anterior segment pathology).
    A. Glaucoma, including narrow angle, open angle, hypersecretion, secondary to increased episcleral venous pressure and congenital.

    X. Trauma.
        A. Contusion or avulsion of the optic nerve.
        B. Subarachnoid hemorrhage of optic nerve.
   XI. Metabolic disorders.
        A. Diabetes mellitus.
        B. Hypoparathyroidism.
        C. Hypertension.
  XII. Toxins.
        A. Quinine.
        B. Arsenic.
        C. Salicylates.
        D. Vitamin-A intoxication.
        E. Methyl alcohol.
        F. Tobacco.
 XIII. Miscellaneous.
        A. Arachnoidal adhesions.
        B. Meningeal hydrops (pseudo-tumor cerebri).
        C. Hydrocephalic enlargement of the third ventricle, acting as a mass.

Whenever any of the above-mentioned etiologic agents damages the optic nerve unilaterally, there will be decreased function in the nerve. The signs and symptoms of this dysfunction will in part be dependent upon the nature of the causative agent as well as upon extent and location of the optic-nerve lesion itself. The following signs are typical of optic-nerve lesions:

1. Decreased visual acuity (ipsilateral).
2. Visual field defects, including increased blind spot, peripheral and/or central scotomata, etc.
3. A normal, pale or choked disk seen on ophthalmoscopy. Sometimes an optic pit, drusen or coloboma may be present. In other conditions, papilledema may be evident.
4. Pupils equal in size because of the equal decussation of afferent pupillary fibers (pretectal nuclei) in the posterior commissure of the mesencephalon (see Fig. 17). This observation also precludes that the lesion has not damaged

any of the neighboring sympathetic or parasympathetic outflow pathways.

5. Pupils of equal size in *unilateral amaurosis* (total blindness).

6. In *bilateral amaurosis*, pupils which show amaurotic *mydriasis*.

7. Pupillary light reflex altered on direct stimulation. There is a weak pupillary miosis (ipsilateral) which is comparable to Lowenstein's "low-intensity reaction." More specifically, there is an increased latency and a decreased amplitude and duration to the miosis of the direct-light reflex on the side of the lesion.

8. Consensual light reaction which is decreased or absent when the light stimulus is presented to the ipsilateral eye.

9. An increased latency in the pupillary mydriasis on dark stimulation.

10. Positive swinging flashlight test (ipsilateral). This is a reflection of abnormal pupillary escape and is also termed the *Gunn pupil sign* (p. 61).

11. Near-vision reflexes appear to be intact.

Optic Chiasm Lesions

The optic chiasm is the crossroad of the afferent visual fiber complex that is headed for the lateral geniculate bodies as well as the pretectal nuclei of the mesencephalon. The axons of retinal ganglion cells whose cell bodies are located within the nasal half of each retina cross to the contralateral side of the midline. It is estimated that 50 percent of all the axons within each optic nerve decussate at the optic chiasm. The chiasm has important anatomical relationships with the pituitary gland, internal carotid arteries, cavernous sinuses and third ventricle of the brain. In order to understand and diagnose the various pathological entities that affect the chiasm, it is important to have a good three-dimensional understanding of these anatomical relationships.

Some of the major etiological conditions that are responsible for chiasmal lesions are listed in anatomic order in the outline below.

Differential Diagnosis of Optic Chiasm Lesions

The following are possible causes of optic chiasm lesions:

I. Infrachiasmatic lesions (bitemporal hemianopsia).
   A. Pituitary adenomas, including chromophobe, acidophilic and basophilic.

II. Suprachiasmatic lesions (bitemporal hemianopsia beginning in the inferonasal quadrants).
   A. Anterosuperior.
      1. Meningiomas of the olfactory groove, tuberculum sellae, lesser wing of the sphenoid and frontal lobe.
      2. Gliomas of the frontal lobe.
      3. Aneurysms of the anterior cerebral and anterior communicating arteries.
   B. Posterosuperior (with central hemianopic scotomas secondary to compression of the posterior chiasmatic angle).
      1. Craniopharyngiomas.
      2. Dilatation of the third ventricle.
      3. Cholesteatomas (rare).
      4. Osteomas (rare).
      5. Osteochondromas (rare).

III. Perichiasmatic lesions involving the chiasm (binasal hemianopsia are common, although many are partial or asymmetrical).
   A. Basilar syphilitic meningitis with chiasmatic adhesions.
   B. Traumatic basilar meningitis.
   C. Encephalitis.
   D. Methanol poisoning.
   E. Frontal sinusitis.
   F. Multiple sclerosis.
   G. Basilar sarcoid.
   H. Hypothalamic tumors.
   I. Arachnoid cysts.
   J. Hand-Shüller-Christian disease.

IV. Lesions on the lateral portion of the chiasm (nasal or binasal hemianopsias).

    A. Internal carotid artery aneurysms.
    B. Direct trauma (rare).
  V. Intrachiasmatic lesions (bizarre bitemporal field defects).
    A. Gliomas.
    B. Chiasmatic neuritis as part of a posterior extension of optic neuritis (very rare).

CLINICAL FEATURES OF OPTIC CHIASM LESIONS

The clinical signs of optic chiasm lesions will depend in part upon the nature of the pathology as well as the location and extent of the lesion. However, a list of the clinical signs in this condition is as follows:

  I. Decreased visual acuity.
  II. Hemianopic field defects.
  III. Wernicke's hemianopic pupillary reflex (doubt), which refers to directing a pencil of light either to the temporal or nasal half of the retina and eliciting a different pupillary response (see Fig. 23). However, in most clinical situations the light stimulus is not discrete enough for this test and there is subsequent internal scatter, so that both halves of the retina are stimulated, negating Wernicke's differential response.
  IV. Visual field defects produced by the chiasmal lesion, which are
    A. *symmetrical.* Then you will get a symmetrical reaction of the pupils to light stimulation. Therefore the swinging-flashlight test would be negative.
    B. *asymmetrical.* Then the eye with the smaller visual field defect will respond better (more miosis) to the light stimulation.
  V. In general, no anisocoria. Again this requires that no damage to the sympathetic and parasympathetic outflow tracts has taken place.

Optic-Tract Lesions

The optic tract is a direct continuation of the axons that were in the optic nerve and optic chiasm on their way back to

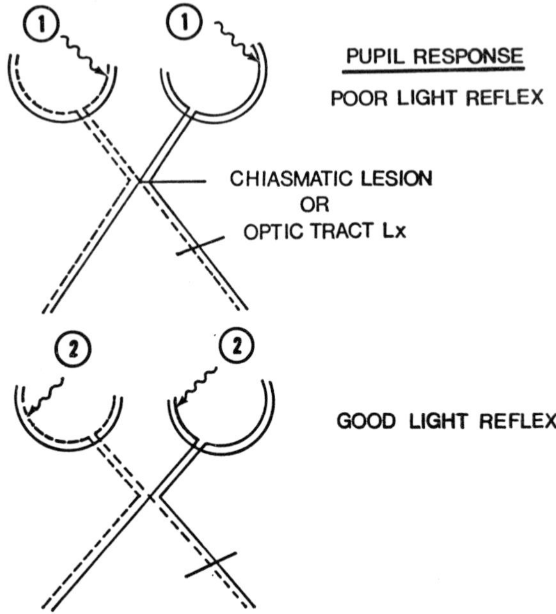

Figure 23. Wernicke's hemianopic pupillary reflex. In lesions of the optic tract or of the optic chiasm, there is a differential response of the pupil (miosis) when the nasal versus the temporal half of the retina is stimulated with light. In the above diagram, with a lesion on the right, light is directed (*1*) toward the side of the lesion and the pupillary response is a weak miosis. But light directed (*2*) away from the side of the lesion evokes a better miosis reaction. However, in most clinical situations, the light stimulus is not discrete enough and there is internal scatter with a loss of the hemianopic pupillary reflex.

either the lateral geniculate bodies or the pretectal nuclei. Again, a thorough knowledge of the important neuroanatomical relationships of nearby structures, i.e. posterior portion of the circle of Willis, frontal lobes and cerebral peduncles, to the optic tract is essential to the understanding and correct diagnosis of optic-tract lesions.

Differential Diagnosis of Optic-Tract Lesions

Listed below, in anatomical order, are some of the etiologic conditions that produce optic-tract lesions.

I.   Anterior optic-tract lesions.
   A. Pituitary tumors (see "chiasmatic lesions")
   B. Posterior chiasmal lesions (see "chiasmatic lesions")
   C. Aneurysms (saccular)
II. Middle optic-tract lesions.
   A. Aneurysms (saccular) of the posterior communicating artery.
   B. Tumors such as infundibulomas, craniopharyngiomas and gliomas.
   C. Demyelinating diseases such as multiple sclerosis and Schilder's disease.
   D. Trauma with skull fracture.
III. Posterior optic-tract lesions similar to those listed under middle optic-tract lesions.

Clinical Features of Optic-Tract Lesions

The clinical signs of optic-tract lesions are dependent upon the nature of the pathology as well as the actual location or site of the lesion. The following features are present in optic-tract lesions:

I.   Normal visual acuity unless the chiasm is involved.
II.   Homonymous hemianopic visual field defects.
III. Optic atrophy, on ophthalmoscopic examination, is a variable finding depending upon the time interval between the initiation of the pathological process and the observation of the disc.
IV.   Usually no anisocoria if visual field defects are homonymous.
V.   Pupil on the side opposite the lesion larger than the pupil on the same side of the lesion if there is an asymmetrical visual-field defect. This is termed *Behr's sign*. When the light stimulus strikes the hemianopic portion of the field, a hypoactivity of the light reflex can be observed (Wernicke's hemianopic pupillary reaction).
VI.   Low-intensity reaction bilaterally to light stimuli by pupils.
VII. Symmetrical opticokinetic responses. (O.K.N. is normal.)
VIII. Associated findings, including

    A. Autonomic nervous system disturbances.
    B. Decreased states of consciousness.
    C. Diabetes insipidus.
    D. Contralateral pyramidal tract signs.
    E. Endocrinopathies, including inappropriate secretion of antidiuretic hormone.
    F. Altered or diminished sense of smell.

Lesions of the Lateral Geniculate Body

Afferent pupillary fibers in the optic tract continue posteriorly toward the pretectal nuclei in the mesencephalon via the superior brachium, thus bypassing the lateral geniculate bodies. The afferent visual fibers go directly from the optic tract to the lateral geniculate bodies where they synapse. Therefore, a lesion of the lateral geniculate bodies will not affect the usual pupillary reflexes to light stimuli. However, a hemianopic field defect will be present.

Lowenstein has shown that lesions of the optic tract can be differentiated from lesions of the anterior portion of the optic radiations. Optic-tract lesions will show abnormal pupillary reflexes, while the optic radiation lesions do not affect the pupillary fibers and so pupillary reflexes are normal.

Lesions of the Superior Brachium

Because of the rarity of a lesion of the superior brachium, it is more of theoretical interest than of any clinical importance. In such a lesion (see Fig. 24), the visual acuity would be normal and no visual-field defects would be found. However, with the appropriate testing equipment, a Wernicke pupillary reflex would be evident as well as a low-intensity reaction to light (bilateral).

Lesions of the Pretectal Nuclei and the Intercalated Neurons

Afferent pupillary fibers come from the optic tract via the superior brachium to synapse with the pretectal nuclei in the mesencephalon. The axons of these pretectal nuclei (intercalated fibers) go to the ipsilateral Edinger-Westphal nuclei (see Fig. 17). Other intercalated fibers cross in the posterior commissure to synapse with contralateral Edinger-Westphal nuclei (see Fig.

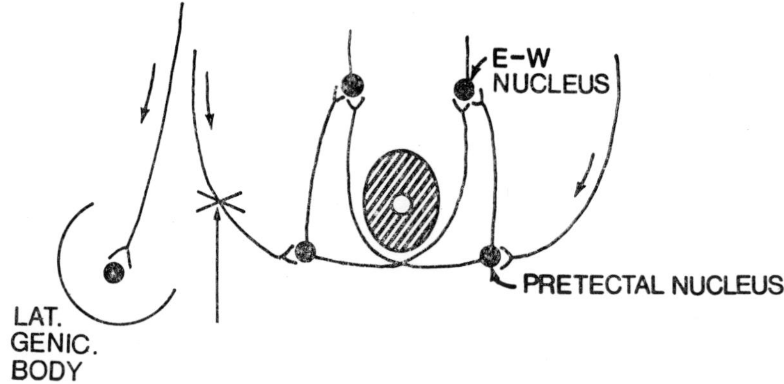

Figure 24. Lesion of the superior brachium. Lesions of the superior brachium would not affect the visual acuity nor the visual field but might bring out Wernicke's hemianopic pupillary reflex.

17). Because of the almost equal decussation of the posterior commissural fibers, the normal direct and consensual light reflexes are equal to each other in timing and intensity.

Alternating Contraction Anisocoria

This is due to asymmetrical decussation of the intercalated neurons in the posterior commissure of the mesencephalon. It is also seen in cases of mild encephalitis. Clinically there is an apparent defect in the consensual light reaction, resulting in an alternating contraction anisocoria. However, this is often a pupillographic diagnosis. Because of the subtle nature of this lesion as well as the fact that a flashlight is used routinely to test pupillary responses to light, alternating contraction anisocoria is often confused with other afferent-arc lesions (see Fig. 25).

Parinaud's Syndrome

Clinical Features

A lesion in the rostral portion of the mesencephalon and located at the posterior commissure level results in *Parinaud's syndrome.* It is usually due to a *pinealoma.* The characteristics of the syndrome are as follows:

1. Loss of the light reaction resulting in dilated pupils.
2. Supranuclear vertical-gaze palsy.
3. Decreased accommodation.
4. Convergence retraction nystagmus on attempted upward gaze.
5. Increased cerebrospinal fluid pressure with papilledema.
6. Intact near reaction (miosis on accommodative effort).
7. No chiasmatic signs.
8. In very rare cases, the pupils which are miotic and fail to react to light stimuli, can be confused with an Argyll Robertson pupil.
9. Cerebellar signs.
10. Extraocular muscle palsies, including ptosis.
11. Deafness.
12. Displaced pupils.
13. Associated precocious sexual development in preadolescent boys in cases of certain pineal tumors.

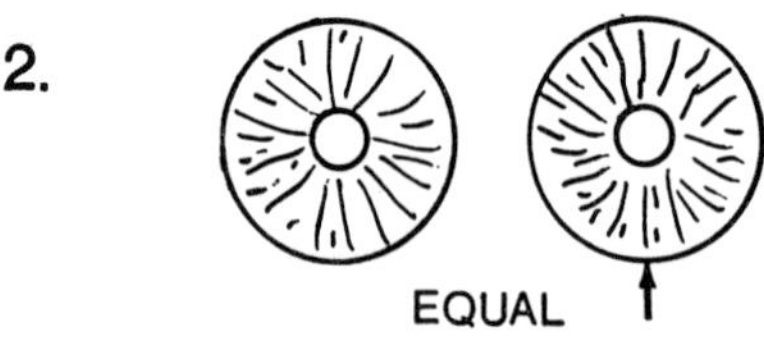

Figure 25. Alternating contraction anisocoria. *Lx* refers to the side of the lesion, which in the above illustration is on the patient's left side. When light stimuli are presented to the contralateral eye, there is a normal, strong, direct pupillary miosis and a weak consensual pupillary miosis, resulting in alternating contraction anisocoria. Often the anisocoria is so slight that this lesion becomes a pupillographic diagnosis.

It should be noted that dilated pupils due to the loss of the light reflex may be the very first sign in Parinaud's syndrome.

Lesions of the Posterior Commissure

A lesion of the posterior commissure does not destroy the consensual light reflex, assuming, of course, that the optic chiasm and other pathways are intact.

Argyll Robertson Pupil

In 1869, Douglas Argyll Robertson reported on "Four cases of spinal miosis: with remarks on the action of light on the pupil" in the *Edinburgh Medical Journal* (15:487-493). For the past 100 years or more, the controversy has raged over where the exact neuroanatomical site of the lesion is in the Argyll Robertson syndrome. Inability to demonstrate a focal lesion on autopsy material as well as failure to reproduce the Argyll Robertson pupil in experimental animals are the main reasons why this controversy has lasted so long. In addition, the definition of the syndrome is disputed by many authors as whether or not to include miosis!

Over the decades, numerous sites have been proposed for the lesion responsible for the Argyll Robertson pupil and have included the following:

1. Cervical cord.
2. Afferent limb of the light-reflex pathway.
3. Ciliary ganglion, short ciliary nerves and iris.
4. Efferent portions of the oculomotor nerve.
5. Pretectal oculomotor fibers.

Although it is by no means conclusive, recent evidence points to the site of lesion in the Argyll Robertson pupil as being in the region of the Sylvian aqueduct in the rostral mesencephalon. The lesion interferes with the light-reflex (afferent arc information) fibers or the intercalated fibers as well as the supranuclear inhibitory fibers as they approach the Edinger-Westphal nucleus (see Fig. 26). This lesion does not affect the descending accommodative fibers that synapse with the Edinger-Westphal nucleus.

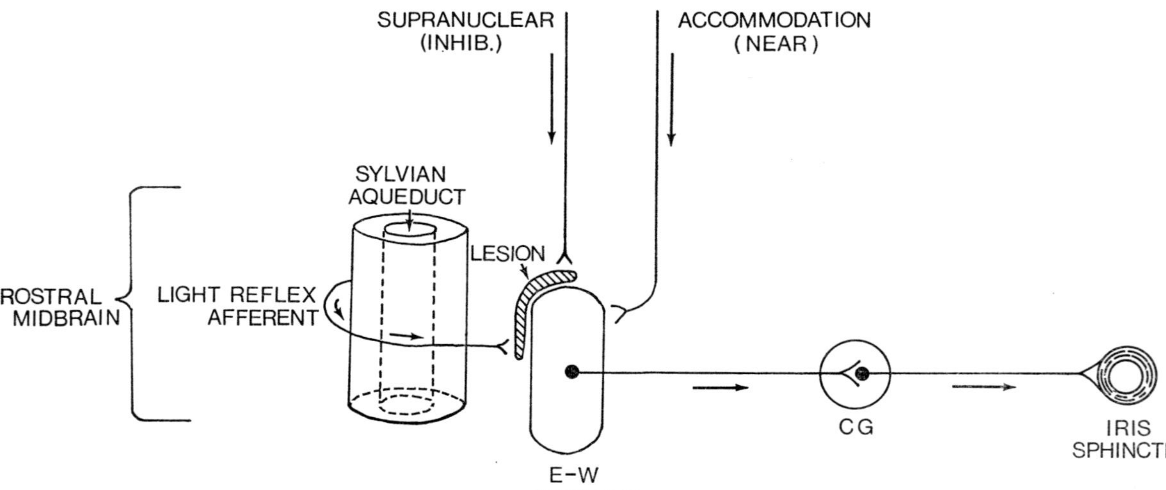

Figure 26. Argyll Robertson pupil: site of neuroanatomical lesion. Note that the proposed site of the Argyll Robertson pupil lesion affects the transmission of light-reflex afferents as well as descending supranuclear inhibitory fibers to the Edinger-Westphal nucleus (*E-W*). Descending accommodation fibers, which are located in the ventral or anterior portion of the mesencephalon, are unaffected by the lesion.

Because of the wealth of literature available on the Argyll Robertson pupil, some of which is rather confusing, it would be worthwhile to devote some time to the nomenclature and definitions regarding this syndrome.

Arygll Robertson sign refers to a pupil that is greater than 2.5 mm in diameter (nonmiotic pupil) that does not respond to light stimuli, yet will undergo miosis on near or accommodative effort. These pupils are seen in cases of orbital trauma with peripheral nerve injuries. Some authors have included *Argyll Robertson sign pupils* in their series with true Argyll Robertson syndromes (defined below), and this has led to considerable confusion. Also "Argyll Robertson sign" is not a good term because it implies that the patient has neurosyphilis when this is far from the case.

CLINICAL FEATURES OF THE ARGYLL ROBERTSON PUPIL OR SYNDROME

The following are clinical features of the Argyll Robertson pupil:

1. Begins as a series of small changes in pupillary function which in the early phases has been termed incipient Argyll Robertson pupil. This eventually progresses to the classical Argyll Robertson syndrome whose features are listed below.
2. The miotic pupil with a diameter of less than 2.5 mm.
3. Visual acuity which is not significantly impaired.
4. Usually bilateral. There are some case reports of unilateral Argyll Robertson pupil, but usually on closer examination, the second pupil is also involved.
5. Irregular contour of each pupil; no bilateral symmetry. Therefore anisocoria or dyscoria are frequently present.
6. Abolished direct and consensual light reflexes.
7. Brisk near-reaction or miosis on accommodation except at the end stage of the disease where the sphincter muscle becomes atrophied and fixed to all stimuli. However, in the earlier stages of the disease, the near-reaction may even be increased.

8. Poor or absent ciliospinal reflex to pain. Yet other psycho-sensory stimuli may or may not evoke mydriasis.
9. No mydriasis evoked by dark stimulation, even after a one-hour period.
10. Poor pupillary response to atropine. However, in the absence of iris damage, Argyll Robertson pupils will dilate with atropine or cocaine and will constrict with miotics, i.e. pilocarpine.
11. Iris atrophy with loss of radial folds and crypts. In some cases, sector atrophy of the iris is present, leading to irregularly shaped pupils. Ovoid depressions in the super-ficial stroma as well as an eccentrically placed vasculosus iridis minor have been observed. However, it is important to note that some patients with lues and iris atrophy do not have Argyll Robertson pupils.
12. Irregular pigment deposition in the iris as well as gross heterochromia.

It should be noted that once the Argyll Robertson pupil has developed, it may remain stationary or go on to a fixed miotic pupil. Also these patients may develop a *luetic ophthalmoplegia* with complete loss of accommodation. Mydriatic pupillary im-mobility may be noted.

The vision in the eye must be sufficient to enable the eye to react to light. If one eye is blind and it does not react to light directly as expected, you cannot make a diagnosis of Argyll Robertson pupil. On the other hand, if both eyes are blind, it is impossible to make the Argyll Robertson pupil diagnosis.

One must be careful in testing a blind patient's pupils. For example, they may not react to light either in the direct or consensual manner, yet may have a near or accommodative miosis. How can this occur? If the flashlight switch makes a noise, the patient may make a near or accommodative effort himself with pupillary miosis occurring, giving the examiner the false impression that this is an Argyll Robertson pupil.

Differential Diagnosis of the Argyll Robertson Pupil with Light- and Near-Reflex Dissociation

The following are possible causes of the Argyll Robertson pupil:

1. Neurosyphilis (the most common cause).
2. Tabes diabetica.
3. Hereditary amyloidosis.
4. Adie's tonic pupil (this pupil is nonmiotic).
5. Aberrant regeneration of the IIIrd nerve (miotic, pseudo-Argyll Robertson pupil).
6. Argyll Robertson sign pupils (often secondary to trauma with damage to peripheral nerves).
7. Central nervous system neoplasms
   a. Pinealomas (Parinaud's syndrome) pupils (nonmiotic).
   b. Midbrain tumors (large pupils).
   c. Posterior third ventricle tumors.
8. Herpes zoster ophthalmicus. (There is no miosis, but there is a peripheral iris lesion.)
9. Trauma to the orbit (similar to the Argyll Robertson sign).
10. Chronic alcoholism (miosis).
11. Multiple sclerosis.
12. Senile and degenerative lesions of the central nervous system, senile dementia.
13. Encephalitis (miosis).

The classical Argyll Robertson pupil is invariably caused by neurosyphilis. Therefore, when the differential diagnosis comes up, it is important to pursue the possibility of lues by ordering VDRL and, if necessary, an FTA-ABS. In some patients with neurosyphilis who are seronegative, a treponema immobilization test may be positive. Anterior chamber taps with subsequent dark-field examination, as stressed by J. L. Smith, are also very important.

Altered Inhibition of the Edinger-Westphal Nucleus

Normally there are corticotectal fibers that impinge upon the Edinger-Westphal nucleus with inhibitory impulses. If, for some

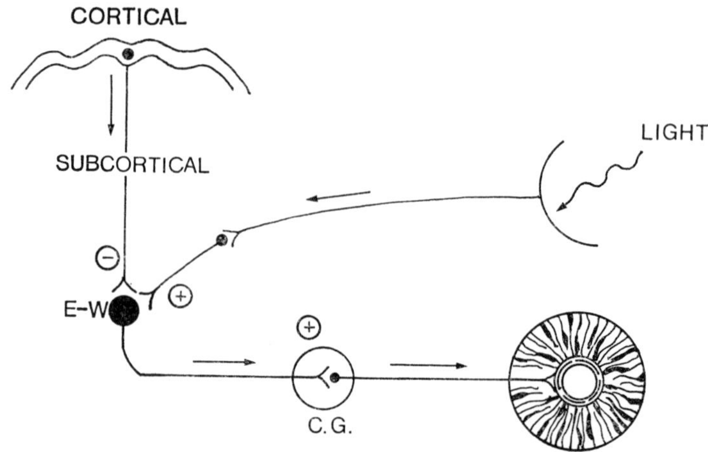

Figure 27.   Defective inhibition of the Edinger-Westphal nucleus.   Decreased cortical inhibitory impulses to the E-W nucleus results in miotic pupils that dilate poorly in the dark, yet have a brisk reaction to light stimuli. Plus (+) indicates stimulation and minus (−) indicates inhibition.

reason, these inhibitory impulses are blocked or diminished, the pupil will become miotic (see Fig. 27). In addition, there will be a poor mydriatic response with dark stimulation. The direct and consensual light reactions are intact. The miosis is, however, from a "parasympathetic disinhibition" and can be relieved by anticholinergic agents, i.e. homatropine. The defective inhibition of the Edinger-Westphal nucleus is found in fatigue, sleep, general anesthesia and in coma. It has also been termed the "miosis of fatigue."

Spastic Miosis

Spastic miosis is usually a bilateral condition often associated with a total or partial loss of the light reaction. The near reaction of the pupil is intact. The condition is comparable to spastic paresis of an upper-motor-neuron lesion. Spastic miosis is seen in the following conditions:

1. Arteriosclerotic and degenerative diseases of the cerebrum.
2. Alcoholism.
3. Myotonic dystrophy.
4. Long-standing diabetes mellitus.

CLINICAL FEATURES OF SPASTIC MIOSIS

The following are clinical features of spastic miosis:

1. Anisocoria.
2. Miotic pupils (usually bilateral).
3. Partial or total loss of direct and consensual light reaction.
4. Dark stimulation not resulting in mydriasis.
5. Only a moderate mydriasis given by anticholinergic agents.
6. Affected near-vision miosis.

Lesions of the Efferent Arc

The efferent arc of the parasympathetic pupillary pathway consists of a two-neuron chain, (a) the Edinger-Westphal nucleus and its axons (*preganglionic fibers*) and (b) the preganglionic fibers synapsing in the ciliary ganglion with *postganglionic fibers* innervating the sphincter pupillae.

CLINICAL FEATURES OF INTERNAL OPHTHALMOPLEGIA

In general, paralysis of the efferent pupillary pathway results in the following clinical signs:

1. Widely dilated pupil (ipsilateral).
2. Pupil which is nonreactive to the direct and consensual light stimuli.
3. Pupil which is nonreactive with near or accommodative effort.
4. Pupil which is reactive to parasympathomimetic agents as well as other miotics.
5. Supersensitivity reaction to mecholyl.
6. Cycloplegia or paralysis of accommodation.

CLINICAL FEATURES OF EXTERNAL OPHTHALMOPLEGIA

Because of the close proximity of the oculomotor complex to the Edinger-Westphal nuclei and the fact that the axons of these nuclei travel together in the trunk of the third cranial nerve as it exits the mesencephalon, it is not common to get isolated pupillary abnormalities. More often there is an accompanying oculomotor fiber damage resulting in external as well as internal ophthalmoplegia. The clinical features are as follows:

1. Ptosis (ipsilateral).
2. Inability to rotate the globe superiorly, inferiorly or medially.
3. Iridoplegia (internal ophthalmoplegia may be present).
4. Cycloplegia (paralysis of accommodation may be present).

DIFFERENTIAL DIAGNOSIS OF IIIRD-NERVE NUCLEAR LESIONS

The following are possible causes of IIIrd-nerve nuclear lesions:

   I. Vitamin deficiencies.
      A. Thiamine (Wernicke's disease).
  II. Toxins and poisons.
      A. Diphtheria (affects accommodation).
      B. Botulism (affects pupillimotor function and/or accommodation).
      C. Dinitrophenol poisoning.
      D. Carbon disulfide (CS_2) poisoning.
 III. Infections.
      A. Amebic dysentery.
      B. Anterior poliomyelitis.
      C. Encephalitis.
         1. Mumps.
         2. Rabies.
         3. Herpes zoster.
         4. Encephalitis lethargica.
         5. Polio.
      D. Hepatitis.
      E. Smallpox vaccination.
      F. Dengue.
 IV. Tumors or neoplasms.
  V. Vascular causes.
      1. Hemorrhages or embolic infarcts of the mesencephalon.
         a. Basilar-vertebral thrombosis.
         b. Basilar embolism.
         c. Primary hypertensive intracerebral hemorrhage.
         d. Mesencephalic hemorrhages secondary to herniation syndromes.

    e. Bleeding disorders including anticoagulation.
    f. Developmental Mobius' syndrome.

DIFFERENTIAL DIAGNOSIS OF IIIRD-NERVE LESIONS WITHIN THE MESENCEPHALON

The causes are similar to those listed for IIIrd Nerve nuclear lesions. Some others are as follows:

1. Benedikt's syndrome, a third nerve lesion that also damages the red nucleus resulting in ipsilateral third nerve paralysis plus contralateral intention tremor.
2. Weber's syndrome is a lesion of the mesencephalon with ipsilateral paralysis of the third nerve as well as damaging the pyramidal tracts (cerebral peduncle) giving contralateral hemiparesis.

DIFFERENTIAL DIAGNOSIS OF IIIRD-NERVE LESIONS FROM THE BRAIN STEM TO THE APEX OF THE ORBIT

The following are causes of IIIrd-nerve lesions from the brain stem to the apex of the orbit:

I. Infection.
    A. Syphilis at the base of the brain.
    B. Tuberculosis at the base of the brain.
    C. Herpes zoster.
    D. Mumps.
    E. Bacterial meningitis.
    F. Sarcoid.

II. Neoplasms.
    A. Pituitary adenomas.
    B. Nasopharyngeal carcinomas.
    C. Chordomas.
    D. Sphenoidal ridge meningiomas.
    E. Temporal lobe tumors.
    F. Meningeal carcinomatosis.
    G. Leukemia.

III. Herniation of the tentorium cerebeli.
    A. Secondary to tumors as listed above, etc.
    B. Secondary to cerebral edema from migraine headache.

IV. Trauma
   A. Cerebral contusion; laceration with resultant edema.
V. Vascular.
   A. Subdural hematomas (ipsilateral mydriasis).
   B. Rupture or rapid growth of an aneurysm of the posterior cerebral or posterior communicating artery.
   C. Cavernous sinus thrombosis.
      1. A-V communication with pulsating exophthalmos.
   D. Cerebrovascular accident (stroke).
   E. Diabetes mellitus and other vasculitides.
VI. Toxins.
   A. Alcohol.
   B. Carbon monoxide.
   C. Lead.
   D. Arsenic.

In general, the pupillomotor fibers are more vulnerable than the oculomotor fibers to the effects of pressure and stretching in the region from the brain stem up to the cavernous sinus (see Fig. 28). The reverse appears to be true within the cavernous sinus up to the superior orbital fissure. However, the final clinical features will depend upon which fibers are damaged as well as the location of the lesion, etc.

Lesions affecting the third cranial nerve within the cavernous sinus or apex of the orbit only show minor pupillary mydriasis because there is simultaneous involvement of the sympathetic fibers as well. In fact, there are some cases of superior orbital fissure lesions where the pupillary reactions were reported as normal on routine clinical testing.

Superior Orbital Fissure or Lateral Wall of the Cavernous Sinus Syndrome

This is characterized by the following:
I. Unilateral third nerve paralysis.
   A. External and/or internal ophthalmoplegia.
II. Trochlear nerve (IV) and abducens (VI) which may be involved.
III. Miotic pupil.
IV. Optic nerve involvement.

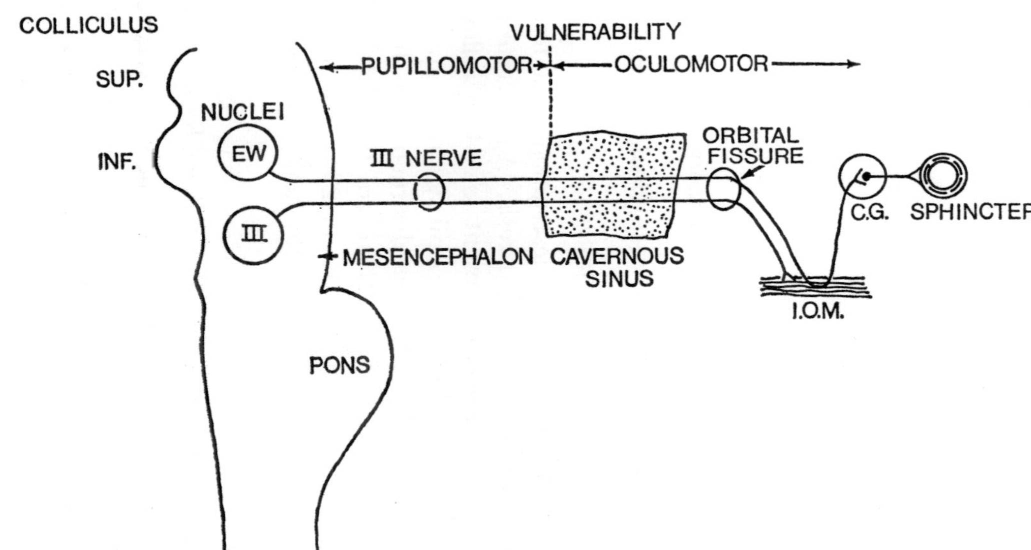

Figure 28. Vulnerability of pupillomotor versus oculomotor fibers. From the brain stem to the cavernous sinus, the pupillomotor fibers are more susceptible to pressure and stretching, whereas the oculomotor fibers are more sensitive to these modalities in the cavernous sinus and orbital regions.

DIFFERENTIAL DIAGNOSIS OF THE SUPERIOR ORBITAL FISSURE SYNDROME

The following are possible causes of the superior orbital fissure syndrome:

I. Inflammation.
   A. Suppuration from the sphenoidal sinus.
   B. Other local inflammation.
II. Neoplasms.
   A. Sphenoidal ridge meningioma.
   B. Nasopharyngeal carcinomas.
   C. Metastatic carcinomas.
   D. Retrobulbar tumors.
   E. Retinoblastoma.
III. Trauma with skull fracture and/or hemorrhage.

Lesions of the Ciliary Ganglion and Postganglionic Fibers

Adie's (Tonic) Pupil

In 1899, Piltz published the first description of a tonically acting pupil in patients with general paresis. Adie wrote about the tonic pupil and the associated absent knee-jerk reflexes in 1931. The primary lesion in the tonic pupil syndrome appears to be a *ciliary ganglionitis* with subsequent destruction of the ganglion cells and their axons which travel toward the globe within the short ciliary nerves (see Figs. 19 and 29). The causative agent of the ganglionitis in most cases of the tonic pupil syndrome is unknown.

DIFFERENTIAL DIAGNOSIS OF THE TONIC PUPIL SYNDROME

The following are possible causes of the tonic pupil syndrome:
I. Idiopathic.
II. Toxins.
III. Mechanical.

   A. Secondary to local tumors.
   B. Metastatic breast carcinoma to the choroid and the posterior extension to the region of the ciliary ganglion.
   C. Trauma.

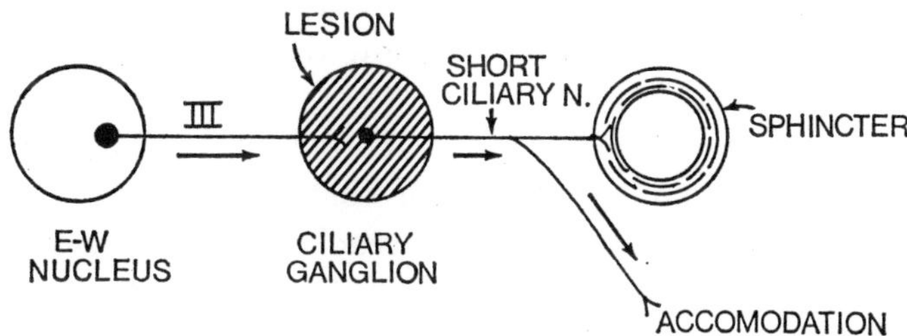

Figure 29.   Adie's (tonic) pupil syndrome: pathology. Destruction of the ciliary ganglion by a variety of causes, the most common of which is idiopathic, results in a tonic or sluggish pupil.

CLINICAL FEATURES OF THE TONIC (ADIE'S) PUPIL SYNDROME
The following are clinical features of the tonic pupil syndrome:

I.   Unilateral in 80 to 90 percent of cases.

II.   Average age of onset between 20 and 40 years.

III.   Females affected more than males (about 3:1).

IV.   In acute stages of the disease, there is an internal ophthalmoplegia with complete loss of pupillary sphincter action as well as accommodation. However, with recovery in the chronic phase, collateral fibers sprout, usually from the accommodative axons, reinnervating the pupil. According to Warwick, 94 percent of synapsing fibers in the ciliary ganglion are accommodative, whereas only 3 percent are involved in pupillomotor functions. Therefore with regeneration of nerves, it is likely that preganglionic fibers will hook up with some of the postganglionic pupilomotor fibers in the repair process. This means that recovery of any accommodative defects will occur much faster than that of pupillomotor defects.

V.   Slow course of the disease. It may go unnoticed, especially if accommodation is uninvolved.

VI.   Pupil on affected side becomes smaller as disease progresses, anisocoria depending on conditions of illumination.
A. When examined in bright light, tonic pupil larger

than normal.

B. When examined in the dark, tonic pupil smaller than normal.

VII. Pupillary reaction to light appearing to be absent. Yet careful observation will show that it is very sluggish and delayed.

VIII. Very sluggish pupillary reaction to near effort. But careful observation will establish that miosis does take place. Once the miotic state is reached and the near effort is discontinued, it will take a long time for the pupil to redilate as contrasted with normals.

IX. Pupil failing to dilate well on dark stimulation.

X. Noticeable vermiform movements of the iris on slit lamp examination. More specifically, various segments of the sphincter muscle appear to be contracting independently in an uncoordinated manner. These can best be seen in patients with blue irides.

XI. Absent or decreased deep tendon reflexes in 50 to 70 percent of patients. These altered reflexes appear to be confined to the knee and ankle. Autopsy studies have shown a progressive degeneration of the local segmental neurophil that supplies the alpha motor neurons in the spinal cord. Another term for the syndrome is pupillotonia pseudotabes.

XII. Often, associated autonomic disturbances, i.e. familial orthostatic hypotension.

XIII. Significant miosis induced by instillation of 2.5 percent mecholyl into the conjunctival sac. This is believed to be due to denervation hypersensitivity. There is no pupillary response when 2.5 percent mecholyl is instilled into the conjunctival sac of normal patients. The 2.5 percent methacholine (mecholyl) test is positive in:

A. Tonic pupil (Adie's Syndrome).

B. Familial dysautonomia (Riley-Day Syndrome).

C. Hereditary amyloidosis.

D. General paresis with fixed, dilated pupils.

E. Tabes dorsalis patients with Argyll Robertson-like pupils.

This test only indicates a hypersensitivity phenomenon at the cholinergic terminals or receptor sites. In addition, *preganglionic parasympathetic nerve lesions do not give a positive mecholyl test.* Finally concentrations of mecholyl up to 15 percent have no effect, when instilled conjunctivally, on the normal pupil.

XIV. Normal reaction to cocaine and adrenalin.

XV. Slow and extensive miotic reaction in cases of pupillotonia, when the first branch of the trigeminal nerve is stimulated, as in touching the cornea. However, other sensory stimuli will evoke mydriasis.

XVI. Possibly, episodes of painful ciliary spasm.

THE CONSCIOUS PATIENT: SYMPATHETIC NERVOUS SYSTEM LESIONS

Horner's Syndrome

In 1869, a Swiss ophthalmologist, Horner, reported the first case of what has been eponymically termed *Horner's syndrome.* Actually, Pourfoir du Petit, in 1727, was the first to produce this syndrome in dogs on an experimental basis.

CLINICAL FEATURES OF HORNER'S SYNDROME

The following are clinical features of Horner's syndrome.

I. Miosis (ipsilateral).

II. Anisocoria.

III. Ptosis (ipsilateral).

IV. Narrowing of the palpebral fissure (ipsilateral).

V. Facial anhydrosis (ipsilateral).

VI. Ocular hypotony (ipsilateral).

VII. Increased amplitude of accommodation (ipsilateral).

VIII. Pigmentary anomalies of iris stroma (ipsilateral).

IX. Unimpaired pupillary reaction to direct and consensual light stimuli as well as near effort.

X. Pharmacologic pupillary responses (ipsilateral).
A. No mydriasis with cocaine test (2 or 5% concentration).

B. Epinephrine test (1:1000 concentration) equivocal mydriasis.

C. Paradoxical pupillary reaction. With intense emotional excitement there is mydriasis (ipsilateral).

Interruption of the sympathetic outflow that reaches the dilator pupillae muscles of the iris may or may not result in a Horner's syndrome. The nature, location and extent of the lesion within the sympathetic pathway will determine, to a large extent, which of the above clinical signs will be present. The etiologic agents that produce Horner's syndrome will be listed in anatomical order in a later section of the manual.

For the present, let us discuss in some detail the main features that comprise Horner's syndrome.

Miosis (Ipsilateral). Interruption of the sympathetic outflow will result in decreased stimulation of the iris dilator muscle (see Figs. 8-13), and therefore a miotic condition is to be expected. Actually this may be seen immediately after the lesion has affected the sympathetic fibers (especially postganglionic lesions). If the lesion is in the preganglionic fibers, pupillary recovery is better than in the postganglionic lesions.

According to several authors, the iris sphincter has some sympathetic innervation which provides peripheral inhibition of the sphincter under certain conditions. Interruption of such a system, as in Horner's syndrome, would contribute to the ipsilateral miosis as described in the preceeding paragraph.

In cervical chain lesions, there can be regeneration of the nerve fibers and you may get recovery of Horner's syndrome in two months, depending upon the nature of the lesion. However, if the lesion destroys the superior cervical ganglion, it is bound to be permanent.

Stimulation of the hypothalamic sympathetic nervous system centers results in ipsilateral mydriasis. This reaction is based on direct stimulation of the dilator pupillae via the sympathetic chain as well as peripheral inhibition of the sphincter pupillae (see Fig. 30) muscle. However, with destruction of the hypothalamic center, there will be an ipsilateral miosis, ptosis and decreased reflex pupillary dilation.

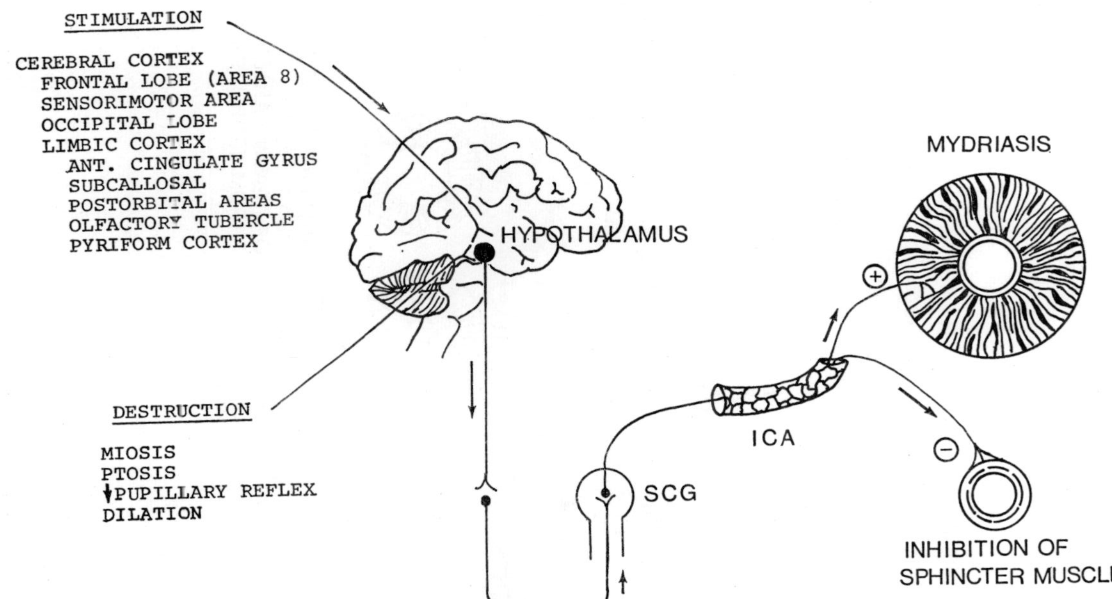

Figure 30. Sympathetic nervous system, hypothalamic center. Stimulation (+) of the hypothalamic sympathetic nervous system center results in mydriasis (ipsilateral) due to direct stimulation of the dilator muscle and to peripheral inhibition (−) of the sphincter muscle. Destruction of the hypothalamic center results in miosis (ipsilateral), ptosis and decreased reflex dilation of the pupil.

Anisocoria. With bright light, anisocoria decreases because of the bilateral miotic reaction of the pupils. However, in darkened conditions, the anisocoria becomes more apparent. When the patient is emotionally excited, the anisocoria increases, and when he is fatigued, drowsy or depressed, the anisocoria is less apparent (see Fig. 31). Anisocoria in Horner's syndrome varies according to the following factors:

1. Completeness of the sympathetic nerve injury.
2. Location of the lesion, i.e. anisocoria is greater with postganglionic lesions than with preganglionic lesions.
3. Alertness or degree of consciousness of the patient.
4. Brightness or darkness of background illumination.
5. Degree of deinnervation hypersensitivity.
6. Fixation of patient at near or at distance.
7. Concentration of circulating neurohumoral agents.

Bilateral, equal psychosensory stimulation, i.e. ciliospinal reflex, may be helpful in bringing out a latent paresis of the dilator muscle.

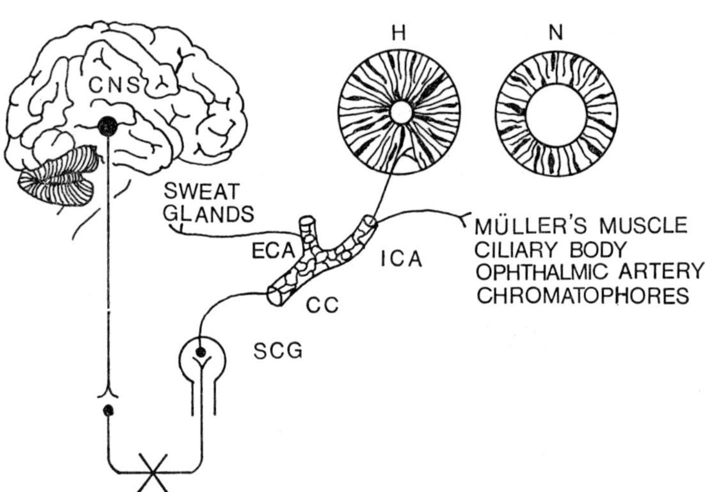

Figure 31. Sympathetic nervous system, anisocoria of Horner's syndrome. Note that a lesion in the cervical sympathetic chain will produce a Horner's syndrome which includes ipsilateral miosis, anisocoria, ptosis (Müller's muscle affected), anhydrosis, etc.

PTOSIS. Ipsilateral ptosis is due to decreased sympathetic supply to Müller's muscle in the upper lid. The lesion also affects the sympathetically innervated muscle of the lower lid (see Figs. 13 and 31).

NARROWING OF THE PALPEBRAL FISSURE. The narrowing of the palpebral fissure is due to the ptosis of the upper lid as well as the elevation of the lower lid (due to decreased sympathetic supply to Müller's muscle fibers in the lower lid). This narrowed palpebral fissure gives the impression of *enophthalmos*. However, exophthalmometer readings are within normal limits.

FACIAL ANHIDROSIS (IPSILATERAL). Ipsilateral facial anhidrosis or decreased sweating, as well as decreased vasoconstriction of the ipsilateral blood vessels to the skin region, are clinically evident. The skin on the face will be warm and dry ipsilaterally. However, the blood vessels (ipsilateral) are more susceptible to circulating neurohumoral agents and so conjunctival hyperemia, epiphoria, nasal stuffiness and flushing of the skin may be observed.

The distribution of anhidrosis is a valuable localizing sign in Horner's syndrome, and it can be mapped out using the following techniques:

1. Starch iodine technique.
2. Quinizarin powder technique where the material is dusted over the patient's skin and he is placed in an overheated room.
3. Measurement of skin temperature electronically.
4. Measuring the electrical skin resistance. When the skin is dry, the electrical resistance increases.

Knowledge of the neuroanatomical course and distribution of the sweat fibers can be a valuable aide in localizing the lesion site in Horner's syndrome. The lesion sites and their abnormal sweat patterns are shown in Table II.

OCULAR HYPOTONY (IPSILATERAL). Ocular hypotony is seen in the early periods after the lesion is established. However, with time there is recovery with a rise in the intraocular pressure back to the normal range.

TABLE II
LOCATING LESION SITE BY SWEAT PATTERN

Site of Lesion	*Sweat Pattern*
Brain stem	Loss of sweating pattern variable, but may include entire ipsilateral half of body including the face.
Ventral roots of cervicothoracic spinal cord.	Anhidrosis of ipsilateral half of face & neck. Occasionally upper extremity also.
Cervical Sympathetic Chain below the bifurcation of int. & ext. carotid art.	Anhidrosis of entire ipsilateral side of the face, especially forehead.
Postganglionic sympathetic fibers in the neck, base of skull or in orbit.	Anhidrosis of ipsilateral forehead may occur.
Sympath. fibers on external carotid art. after bifurcat. from int. carotid artery.	Anhidrosis of ipsilateral face except over nasion region and area between eyebrows (see Fig. 14).

INCREASED AMPLITUDE OF ACCOMMODATION. Increased amplitude of accommodation is found, which may be due to effects on the ciliary body or the lack of pupillary mobility. It has been shown that the latter factor can alter the range of accommodation by 1.5 diopters or more. In addition, the pupil is miotic and the depth of focus is therefore increased, which may contribute to the increased amplitude of accommodation measurement.

PIGMENTARY ANOMALIES OF IRIS STROMA (IPSILATERAL). Iris pigmentary anomalies are seen in congenital Horner's syndrome. It is believed that the sympathomimetic agents are involved in some manner with the normal development of pigment in the iris stroma. Absence of sympathetic innervation alters the normal sequence, and a heterochromia ensues with the lighter iris on the ipsilateral side of the lesion. Onset of Horner's syndrome after one to two years of age rarely produces heterochromia.

PUPILLARY REACTIONS TO DIRECT AND CONSENSUAL LIGHT STIMULI. The pupillary reactions to direct and consensual light stimuli are miotic reactions and do not appear to be impaired in Horner's syndrome. The same is true for the *near or accommodative reaction.*

PHARMACOLOGIC PUPILLARY RESPONSES (IPSILATERAL). In the *cocaine test,* cocaine in 2 or 5% solution is instilled into the lower conjunctival sac. In normals, there is norepinephrine present at the sympathetic nerve terminals (see Fig. 20). Cocaine acts by blocking the reabsorption or uptake of unused norepinephrine at the postganglionic nerve terminal in the region of the receptor site thereby allowing the norepinephrine to stimulate the effector cell again which in this case is the iris dilator muscle. This action results in mydriasis. However, in Horner's syndrome, there are depleted stores of norepinephrine so that the cocaine has little, if any, neurotransmitter substance to work on. As a result, there is no mydriasis in a Horner's pupil with local cocaine administration (see Fig. 32).

However, if the lesion is in the brain stem, its effects on depleting terminal norepinephrine stores at the iris dilator level may be nil. This is the reason why you may get mild mydriasis with cocaine in brain-stem lesions. Also if you use a 10% cocaine solution for this test, you may anesthetize the nerves to the iris sphincter and get a false mydriasis. In general, a positive cocaine test helps to confirm the existence of a Horner's syndrome or pupil but does not clarify the location of the lesion within the sympathetic nervous system.

The *epinephrine test* consists of administering several drops of epinephrine (1:1000) into the lower conjunctival sac bilaterally. This is a very dilute solution and does not affect the normal pupil. However, a Horner's pupil, where there is a depletion of norepinephrine at the nerve terminals and hence a denervation hypersensitivity, should react to the epinephrine and undergo mydriasis. Unfortunately, this does not work consistently nor convincingly in the clinical setting.

Topical administration of cocaine destroys the anterior corneal epithelium, enhancing corneal permeability for epinephrine penetration. In addition, the cocaine sensitizes the sympathetic nerve terminals to adrenergic agents, thereby giving a potentiation effect.

The major problem with the *cocaine test followed by the epinephrine test* is that a normal pupil will dilate with cocaine, but a Horner's pupil will not. Since bilateral comparison of pupil size is important in evaluating a significant pharmacological response, the epinephrine test that follows will be difficult to evaluate. Why is this so? Because, after the cocaine test, you have a

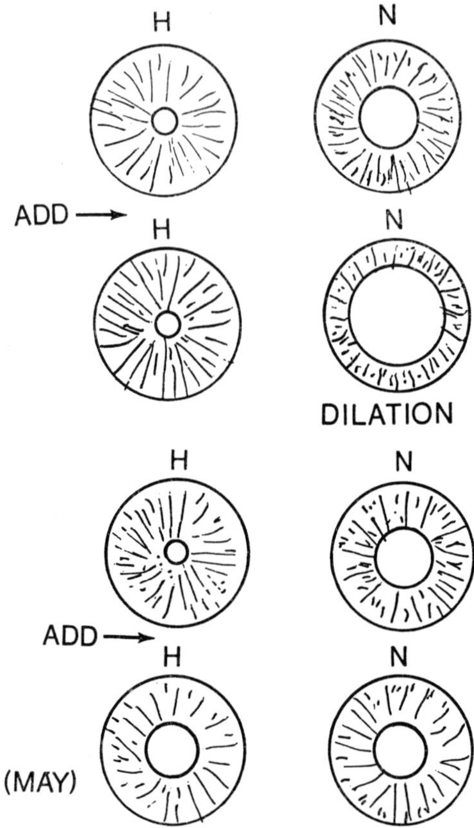

Figure 32. Horner's syndrome, pharmacologic pupillary responses. *Top,* in the cocaine test (2% or 5% solution), a Horner pupil, because of depleted norepinephrine stores at the nerve terminal, will not undergo mydriasis as does a normal pupil. *Bottom,* in the epinephrine test (1:1000 solution), a Horner's pupil may undergo mydriasis because of denervation hypersensitivity. A normal pupil would not react to such a dilute concentration of topical epinephrine. More often than not, the mydriatic response in a Horner's pupil to the epinephrine is equivocal and not of much clinical value.

mydriatic effect in the normal eye and you have destroyed the anterior corneal epithelium bilaterally, giving rise to possible unequal penetration of the epinephrine to follow. Lastly, there is no baseline upon which to evaluate a resultant mydriasis in the Horner's pupil. In addition, the deinnervation supersensitivity varies with the completeness of the lesion and its distance from the iris or nerve terminals, so that it is possible to have a partial postganglionic Horner's pupil with less denervation super-sensitivity than one with a complete preganglionic lesion.

In conclusion, the epinephrine test (1:1000) is not very helpful in the diagnosis of Horner's syndrome nor in delineating the site of the lesion within the sympathetic nervous system.

In cases of Horner's syndrome when there is intense emotional excitement, neurohumoral substances are released into the circulation and find their way to the iris dilator muscle where they act to produce a *paradoxical pupillary mydriasis* (ipsilateral) that may be larger than the mydriasis in the contralateral pupil (Fig. 33).

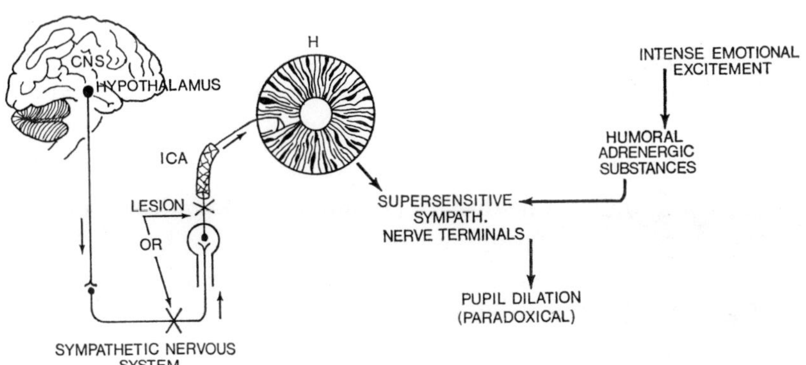

Figure 33.   Horner's syndrome, paradoxical pupillary reaction. In intense emotional states, neurohumoral substances are released into the general circulation and eventually affect the iris dilator muscle (deinnervation hypersensitivity) which in Horner's syndrome dilates paradoxically.

DIFFERENTIAL DIAGNOSIS OF HORNER'S SYNDROME LESIONS
CENTRAL LESIONS OF THE SYMPATHETIC NERVOUS SYSTEM

The following are possible causes of central lesions of the sympathetic nervous system in Horner's syndrome (in anatomic order):

I.   Cerebral lesions. Cerebral lesions have been reported to give contralateral Horner's syndrome. This is supported by experimental evidence as well.

II. Hypothalamic lesions. Lesions of the posterior and lateral nuclei of the hypothalamus will give an ipsilateral Horner's syndrome. There is also experimental evidence to substantiate this finding.

III. Pontine lesions. Although the exact pathway of the descending hypothalamospinal fibers within the pons is not known, certain pontine lesions will give ipsilateral Horner's syndrome.

In cases of pontine hemorrhage and intrapontine tumors there is a pinpoint miosis which is not due so much to the interruption of the descending sympathetic fibers as it is due to the interruption of inhibitory impulses to the Edinger-Westphal nucleus (see Fig. 18). This latter condition allows the Edinger-Westphal nucleus to increase its parasympathetic stimulation of the iris sphincter resulting in miosis.

IV. Medulla. Sympathetic fibers are found in the lateral medulla.

A. Vascular.

    1. Lateral medullary (Wallenberg's) syndrome usually due to thrombosis of the vertebral or basilar artery; less often to thrombosis or embolus of the posterior inferior cerebellar artery.

    2. Syringobulbia.

V. Spinal Cord.

A. Congenital defects in the spinal cord or medulla in association with facial hemiatrophy and scleroderma.

B. Degenerative syringomyelia (15% of cases are associated with an intramedullary hemangioblastoma or glioma).

C. Neoplasms.

D. Demyelinating disorders such as multiple sclerosis.

E. Trauma.

F. Infectious poliomyelitis.

G. Vascular occlusions and vascular malformations.

PERIPHERAL SYMPATHETIC NERVOUS SYSTEM LESIONS
(HORNER'S SYNDROME)

The following are possible causes of peripheral sympathetic nervous system lesions in Horner's syndrome (in anatomic order):

I. Paravertebral lesions.
   A. Trauma.
      1. Birth trauma with stretching of the cervical plexus resulting in *Klumpke's paralysis* and a Horner's syndrome.
      2. Ruptured intervertebral disc.
      3. Cervical vertebral fractures.
      4. Radical neck dissections.
      5. Thyroidectomies.
   B. Neoplasms.
      1. Apical pulmonary tumors such as bronchogenic carcinoma or *Pancoast tumor* (see Fig. 34).
      2. Mediastinal tumors.
      3. Goiter.
   C. Inflammations and Infections.
      1. Apical tuberculosis with cervical vertebral destruction (Fig. 34).
      2. Cervical lymphadenopathy with pressure on the cervical chain.
      3. Pachymeningitis.
      4. Hypertropic spinal arthritis with narrowing of the cervical intervertebral foramina.
      5. Idiopathic brachial neuritis.
   D. Vascular.
      1. Thoracic aortic aneurysms. In the early stages these aneurysms are irritative and therefore stimulate the sympathetic chain (ipsilaterally). In the later, destructive, phases, the sympathetic axons are damaged with Horner's syndrome resulting.
      2. Brachial neuritis of serum sickness.

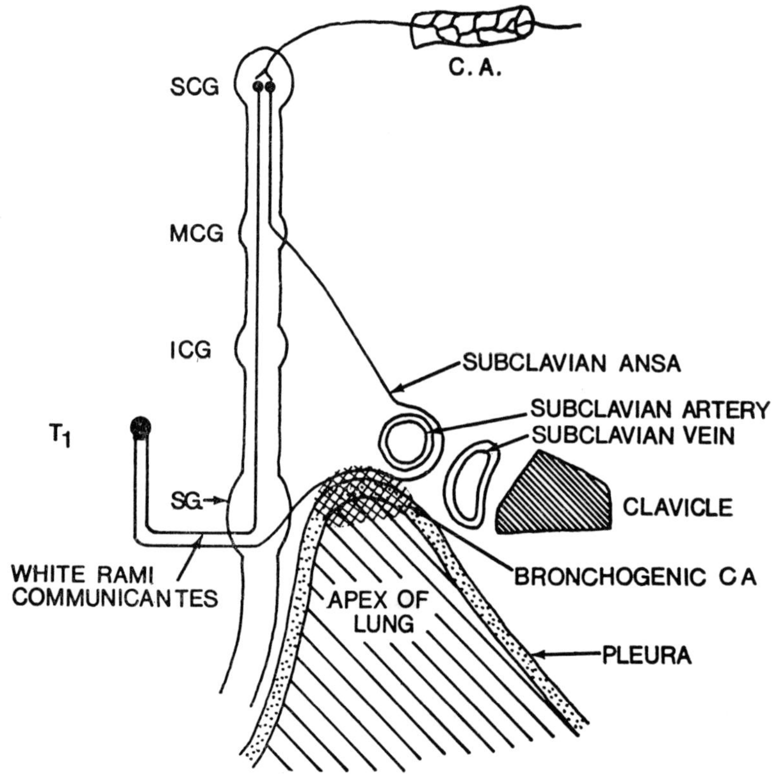

Figure 34. Horner's syndrome, apical pulmonary lesions. Note the anatomical relationship of the sympathetic fibers to the apical lung pleura. Bronchogenic carcinoma (Pancoast's tumor) and apical pulmonary tuberculosis can damage these sympathetic fibers, as illustrated, and produce a Horner's Syndrome. *SG*, stellate ganglion; *ICG*, inferior cervical ganglion; *MCG*, middle cervical ganglion; *SCG*, superior cervical ganglion; *CA*, carotid artery.

II.   Lesions at the base of the skull.
    A. Trauma.
       1. Injuries to the deep retroparotid spaces.
       2. Complicated tonsillectomies.
    B. Inflammations and infections such as a mandibular tooth abscess.
III.   Middle ear lesions.

    A. Inflammations and infections such as suppurative otitis media (see Fig. 11).

    B. Neoplasms such as cholesteatomas.

IV. Intracranial lesions.

    A. Vascular.

       1. Aneurysms of internal carotid artery (*Raeder's paratrigeminal syndrome*).

       2. Carotid artery thromboses.

       3. Cluster headaches or "migrainous neuralgia."

       4. Cavernous sinus pathology (usually results in semi-dilated, fixed pupils with the sympathetic and parasympathetics involved).

    B. Neoplasms. Meningiomas can give a Raeder's paratrigeminal syndrome.

V. Orbital lesions. See page 84.

THE COMATOSE PATIENT

The neurological and ophthalmological examination of the comatose patient are, of course, altered by the condition of the patient and therefore must be tailored to the needs of the clinical situation. Certain portions of the standard examination of conscious patients may have to be changed or omitted, while other features must be scrutinized more closely.

In the examination of the pupils, which would be an integral part of the evaluation of the patient's neurological status, particular attention should be paid to the size, contour and symmetry of the pupils as well as their reaction to light stimuli. Fisher states that before one concludes that the pupil is unreactive to light stimuli, a 100 watt bulb should be shone in both eyes simultaneously, with careful inspection of the pupils for movement, using a hand magnifying lens. The existence of hippus is also important. In addition, one should look for the existence of ocular disease that may affect pupillary movements as the following case demonstrates.

A 78-year-old man in coma was seen in the emergency room. No history was available at that time. Physical examination and general neurological examination provided no localizing signs. Ophthalmological examination revealed the right pupil was semi-dilated, fixed and nonreactive to light stimuli. The consensual light response was intact. The cornea appeared to be hazy. Further examination revealed a rock-hard globe on gross palpation. What is your diagnosis?

In the above case, the patient had an acute attack of glaucoma, but later history revealed that he had right eye pain for approximately one week, during which time he experienced nausea, vomiting and nonspecific anorexia. Because he was a recluse, he declined to seek medical assistance. With progressive dehydration, he became disoriented and finally comatose. A neighbor discovered him in this condition and he was brought to the hospital emergency room where the above findings were noted. Careful examination of this patient, including his eyes, resulted in the correct diagnosis of acute glaucoma.

Many of the lesions that produce coma with abnormal pupillary movements have been discussed in earlier sections of the manual. A few of the conditions that have not been specifically treated as yet are loss of function of various portions of the central nervous system and loss of the pupillary light reflex.

Loss of Function in the Central Nervous System

In 1966, Plum and Posner described the orderly progression of the loss of function of various portions of the central nervous system (diencephalon, mesencephalon, pons and medulla, in that order) as the damage increased from supratentorial mass lesions. In the early phases of this progression, the clinical picture takes either one of two routes, namely a *central syndrome* or the *uncal syndrome*.

Central Syndrome

The central syndrome is a downward herniation of the diencephalon through the tentorial notch. It is seen in rapidly advancing compressions such as are found in intracerebral hemorrhages, acute cerebral edema, subdural hemorrhages and extra-dural hemorrhages. The clinical features of the central syndrome are as follows:

1. Coma.
2. Cheyne-Stokes respiration.
3. Miotic pupils (less than 2 mm in diameter).
4. Extraocular movements and reflexes are intact.
5. There are motor problems in the extremities which are found bilaterally.

Uncal Syndrome

The uncal syndrome is a medial herination of the medial temporal lobe through the tentorial notch. It is seen in brain abscesses, subacute cerebral edema and chronic subdural hematomas (these take two to four days to develop). The clinical features of the uncal syndrome are as follows:

1. Coma.
2. Normal respiratory movements.
3. Mydriatic pupil (6 to 10 mm), ipsilateral.
4. Extraocular movements and reflexes altered as oculomotor paralysis progresses.

Table III is a list of cerebrovascular accidents in various parts of the central nervous system and their effects on the pupillary reflexes, etc.

TABLE III

CEREBROVASCULAR ACCIDENTS

Site of Cerebrovascular Accident	*Alterations in the Pupil*
Supratentorial subdural hematoma.	Slightly miotic pupils (ipsilateral) Ptosis (ipsilateral).
Middle cerebral artery territory infarction; internal carotid artery thrombosis.	Miotic pupil (ipsilateral).
Pontine hemorrhage (early).	Miotic pupils that are pinpoint (1 mm); may be fixed to light.
Pontine hemorrhage (late).	Mydriatic pupils; may be fixed to light (probably secondary to midbrain hemorrhage).
Cerebellar hemorrhage.	Pupils are 2 to 3 mm in diameter and reactive to light.
Basilar artery territory infarction.	Pupils may be dilated and fixed or pinpoint, depending on site of damage.
Basal ganglia (putamenal) hemorrhage.	Average-size pupils or miotic.
Thalamic and subthalamic hemorrhage.	Average-size pupils or miotic.
Bilateral thalamic lesions.	Mydriatic pupils that are fixed.
Mesencephalon hemorrhage.	Huge mydriatic pupils.
Subthalamic hemorrhage.	Miotic pupils.
Midbrain-subthalamic hemorrhage.	Average-size pupils.
Midbrain-pontine hemorrhage.	Average or intermediate-size pupils.

Loss of the Pupillary Light Reflex

Bilateral loss of the pupillary light reflex is seen in thalamic-subthalamic lesions as well as extensive damage to the mesencephalon. The latter condition is usually secondary to brain-stem compression involving the tentorium. Also, small Duret hemorrhages which involve the mesencephalic tegmentum can cause bilateral loss of pupillary light reflex.

Unilateral loss of the pupillary light reflex is seen with acute massive brain-stem compressions. The uncal syndrome is a pertinent feature here. It is interesting to note that in sleep (miotic pupils) and in the Horner's syndrome, the pupillary reflex to light stimulation is still intact.

For further reading on this interesting subject of the comatose patient, the reader is referred to Professor Walsh's textbook on *Clinical Neuro-Ophthalmology* as well as C. M. Fisher's monograph entitled *The Neurological Examination of the Comatose Patient*. Finally, *The Diagnosis of Stupor and Coma*, by F. Plum and J. B. Posner, should also be consulted.

BIBLIOGRAPHY

AT THE PRESENT time, the literature that is available on the pupil is enormous. Because the main purpose of this manual is to serve as an introductory guide to pupillary behavior, at the clinical level, and not a review of all the literature, the reference list will be relatively concise. The reader is encouraged to use this bibliography as a base upon which to build his knowledge and understanding of pupillary behavior. This latter purpose is best accomplished by reading widely on the subject and carefully evaluating authors' conclusions and experimental techniques and comparing them, when applicable, to your own clinical experience.

Adie, W. J.: Complete and incomplete forms of the benign disorder characterized by tonic pupils and absent tendon reflexes. *Br. J. Ophthalmol.*, 16:449, 1932.

Adler, F. H.: *Physiology of the Eye*, 4th ed. St. Louis, Mosby, 1965.

Bender, M. (Ed.): *The Oculomotor System*. New York, Harper and Row, 1964.

Carpenter, R. L.: *Notes for the Laboratory Study of the Normal Histology of the Eye and Associated Structures*. A manual prepared for Harvard Basic Science Course in Ophthalmology, no date.

Cogan, D.: *Neurology of the Ocular Muscles*, 2nd ed. Springfield, Thomas, 1956.

Cogan, D.: *Neurology of the Visual System*. Springfield, Thomas, 1966.

Davson, H.: *The Physiology of the Eye*, 2nd ed. Boston, Little Brown, 1963.

Drews, R. C.: The Gunn pupil sign, *Am. J. Ophthalmol.*, 54:1109, 1962.

Duke-Elder, W. S. (Ed.): *System of Ophthalmology*, Vol. II, *The Anatomy of the Visual System*. London, Henry Kimpton, 1961.

Duke-Elder, W. S. (Ed.): *System of Ophthalmology*, Vol. III, part 2: *Congenital Deformities*. St. Louis, Mosby, 1963.

Duke-Elder, W. S. (Ed.): *System of Ophthalmology*, Vol. VII: *The Foundations of Ophthalmology*. London, Henry Kimpton, 1962.

Enoch, D. M., and Kerr, F. W. L.: Hypothalamic vasopressor and vesico-pressor pathways. I. Functional studies. *Arch. Neurol.*, 16:290, 1967.

Fisher, C. M.: The neurological examination of the comatose patient. *Acta Neurol. Scand.*, 45 (suppl. 36), 1969.

Grant, W. M.: *Toxicology of the Eye*. Springfield, Thomas, 1962.

Gunn, M.: Discussion of retro-ocular neuritis. *Lancet,* 2:412, 1904.

Horner, J. F.: Uber eine Form von Ptosis. *Klin. Monatsbl. Augenheilkd.,* 7:193, 1869.

Kerr, F. W. L.: The pupil—functional anatomy and clinical correlation. In Smith, J. L. (Ed.): *Neuro-Ophthalmology.* St. Louis, Mosby, 1968, Vol. IV, p. 49.

Kestenbaum, A.: *Applied Anatomy of the Eye.* New York, Grune and Stratton, 1963.

Kestenbaum, A.: *Clinical Methods of Neuro-Ophthalmologic Examination,* 2nd ed. New York, Grune and Stratton, 1961.

Levatin, P.: Pupillary escape in disease of the retina or optic nerve. *Arch. Ophthalmol.,* 62:768, 1959.

Last, R. J. (Ed.): *Eugene Wolff's Anatomy of the Eye and Orbit,* 6th ed. Philadelphia, Saunders, 1968.

Lowenfeld, I. E.: Mechanisms of reflex dilation of the pupil. *Doc. Ophthalmol.,* 12:185, 1958.

Lowenfeld, I. E., and Thompson, H. S.: The tonic pupil; a re-evaluation, *Am. J. Ophthalmol.,* 63:46, 1956.

Lowenstein, O.: The Argyll Robertson pupillary syndrome; mechanism and localization. *Am. J. Ophthalmol.,* 42:105, 1956.

Lowenstein, O.: Clinical pupillary symptoms in lesions of the optic nerve, optic chiasm and optic tract. *Arch. Ophthalmol.,* 52:385, 1954.

Lowenstein, O., and Lowenfeld, I. E.: Electronic pupillograph; a new instrument and some clinical applications. *Arch. Ophthalmol.,* 59:352, 1958.

Lowenstein, O., and Lowenfeld, I. E.: Pupillotonic pseudotabes; a critical review of the literature. *Surv. Ophthalmol.,* 10:129, 1965.

McNealy, D. E., and Plum, F.: Brainstem dysfunction with supratentorial mass lesions. *Arch. Neurol.,* 7:10, 1962.

Smith, J. L., and Israel, C. W.: Spirochetes in the aqueous humor in seronegative ocular syphilis. *Arch. Ophthal.,* 77:474, 1967.

Walsh, F. B., and Hoyt, W. F.: Clinical Neuro-ophthalmology, 3rd ed. Baltimore, Williams & Wilkins, 1969, vol. 1.

Warwick, R.: The ocular parasympathetic nerve supply and its mesencephalic sources. *J. Anat.,* 88:71, 1954.

Zuckerman, J.: *Diagnostic Examination of the Eye,* 2nd ed., Philadelphia, Lippincott, 1964.

CONDITIONS AND PHARMACOLOGIC AGENTS THAT AFFECT PUPILLARY SIZE

Anisocoria. Inequality of 1mm or greater between pupils.
   Aneurysm of the aorta or carotid artery
   Cerebrovascular accidents
   Amaurotic eye (blind)
   Wernicke's hemianopic pupil
   Adie's (tonic) pupil
   Spastic miosis
   Horner's syndrome
   Pontine lesions
   Aberrant third nerve regeneration
   Third nerve lesions
   Ocular trauma
   Alternating contraction anisocoria
   Encephalitis (mild cases)
   Normal variation (small percentage of the population)
   Cervical rib (ipsilateral constricted pupil)
   Tabes dorsalis
   Unilaterally instilled pharmacologic agents (miotics or
      mydriatics)
   Glaucoma
   Glass eye
   Iritis
   Trigeminal neuralgia
   Tournay's pupillary reaction
Hippus. When pupillary excursions exceed those of normal
      physiologic pupillary unrest; it has no localizing sig-
      nificance.

 Normals
 Epilepsy
 Meningitis (acute)
 Disseminated sclerosis
 Cerebral tumors
 Incipient cataracts
 Neurasthenia (nervous exhaustion, Beard's disease)
 Barbiturate poisoning
 Paraldehyde poisoning
Miosis. Pupil becomes smaller.
 Pharmacologic agents
  Parasympathomimetics (See Appendix 3.)
   Acetylcholine
   Pilocarpine
  Sympatholytics (See Appendix 3.)
   Ergotamine
   Hydralazine
  Histamine
  Morphine
  Picrotoxin
 Physiologic conditions
  Sleep
  Fatigue
  General anesthesia, Stage III, Planes 1 and 2
  Near reaction
  Direct light reaction
  Consensual light reaction
  Westphal-Piltz phenomenon or forced closure of eyelids (orbicularis)
  Infancy
  Old age (iris atrophy)
  Congenital miosis with absence of dilator muscle
  Tournay's reaction (adducting eye)
  Optomotor field (frontal lobe) stimulation
  Occipital lobe stimulation
 Pathologic conditions
  Areflexia of diabetes mellitus

Argyll Robertson pupil
Alcoholism
Apneic period of Cheyne-Stokes respiration
Amaurosis (blind eye)
Adie's syndrome when there is stimulation of the fifth cranial
    nerve, i.e. corneal irritation.
Aortic aneurysm (late stage) and other thoracic aneurysms
Brachial plexus trauma with Klumpke's paralysis
Apical lung tumors, i.e. bronchogenic carcinoma (Pancoast's
    tumor)
Apical lung infections, i.e. tuberculosis
Coma
Central syndrome (bilateral miosis, reactive pupils)
Cervical rib (ipsilateral miosis)
Cerebellar hemorrhage (miosis, pupillary reflexes intact)
Carbon dioxide poisoning
Cervical lymphadenopathy
Congenital spinal cord defects (associated with scleroderma
    and facial hemiatrophy)
Basal ganglia hemorrhage
Complicated tonsillectomy
Goiter
Dilated iris blood vessels
Encephalitis
Hypertrophic spinal arthritis
Injuries to the deep retroparotid spaces
Migraine headaches (cluster headaches)
Multiple sclerosis
Mediastinal tumors
Pachymeningitis
Miotic rigidity of the iris secondary to long term miotics, i.e.
    pilocarpine, phospholine iodide or DFP
Horner's syndrome (See pp. 97-101 for complete listing.)
Adie's (tonic) pupil, chronic long-term stage (miosis).
Meningitis
Pontine hemorrhage (pupillary reflexes may be intact)

Intraventricular hemorrhage (secondary irritation evokes miosis)
Spastic miosis (bilateral)
Ruptured intervertebral disc, cervical
Vascular malformations of cervical spinal cord
Lateral medullary (Wallenberg's) syndrome (occlusion of posterior inferior cerebellar artery)
Iritis (ipsilateral)
Syringomyelia
Poliomyelitis
Subthalamic lesions
Radical neck dissections
Thyroidectomies
Mandibular tooth abscesses
Raeder's paratrigeminal syndrome

Mydriasis. Pupil becomes larger.
  Pharmacologic agents
    Adrenalin (See Appendix 3.)
    Atropine (See Appendix 3.)
    Benzedrine®
    Cocaine
    Ephedrine
    Ammonium chloride
    Steroids (local application results in mild mydriasis, mechanism unknown)
    LSD-25
    Glutethimide
    Artane®
    Cogentin®
    Nalorphine (reverses miosis of morphine)
    Low serum calcium
    High serum magnesium
  Physiologic conditions
    Fright reaction
    Darkness
    Infancy
    Old age

Myopia
Psychosensory reflexes
  ciliospinal
  cochleopupillary
  vestibulopupillary
  Flateau's neck mydriasis
  Redlich's phenomenon
  Myer's iliac phenomenon
General anesthesia, Stages I and II (secondary to emotional stress and stage II is reflexic mydriasis)
Pathologic conditions
  Unilateral mydriasis
    Adie's pupil (acute phase)
    Extradural hematoma
    Carotid artery occlusion leading to iris atrophy with a fixed dilated pupil
    Coma
    Uncal syndrome (herniation of temporal lobe)
    Downward displacement of the posterior cerebral artery
    Distortion of the brain stem
    Subdural hematoma
    Irritative lesions (ipsilateral mydriasis)
      Apical pulmonary tuberculosis (Pancoast tumor)
      Bronchogenic carcinoma (apical)
      Bronchiectasis
      Postartificial pneumothorax
      Postthyroidectomy
      Aneurysm of aortic arch (early irritative stimulation)
    Midbrain lesions, i.e. hemorrhages
    Midbrain tumors
    Third-nerve palsies
    Glaucoma
    Aberrant regeneration of third-nerve fibers
    Epileptic seizure (intact pupillary light reflexes)
    Tournay's reaction (abducting eye)
  Bilateral Mydriasis
    Diphtheria toxin (postdiphtheritic state)

    Botulism
    Ciliary ganglionitis (retrobulbar neuritis)
    Catatonia pupillary immobility (schizophrenics occasionally have "spasmus mobilis")
    Uncal syndrome
    Painful stimuli of comatose patient
    Paresis
       generalized
       juvenile
    Parinaud's syndrome (pinealoma)
    Hysteria
    Bumke's anxiety pupils
    General anesthesia, Stage IV, paralytic dilatation
    Amaurotic mydriasis (bilateral blindness)

Fixed Pupils.
  Cavernous sinus lesions
  Retro-orbital lesions
  Acute glaucoma
  Parinaud's syndrome
  Iris atrophy secondary to carotid artery occlusion
  Miotic rigidity of the iris secondary to long-term pilocarpine, etc.
  Barbiturates
  Iris bombe with 360° of posterior synechiae
  Acute iridoplegia secondary to trauma
  Rollet's syndrome
  Weber's syndrome
  Bilateral thalamic lesions
  Early pontine hemorrhage (miotic)
  Late pontine hemorrhage (mydriatic)

Normal-Sized Pupils.
  Metabolic coma
  Cerebral infarctions
  Encephalitis
  Anoxia
  Normals

TOXIC AGENTS THAT PRODUCE
MYDRIASIS OR MIOSIS

Aconite is an ancient drug derived from a plant that can produce yellow vision as well as *mydriasis.*

Barracuda meat poisoning can produce *mydriasis.*

Botulinus toxin, produced by *Clostridium botulinum,* interferes with the release of acetylcholine from the ends of the nerve terminals. The receptor or effector-cell site is not affected by the toxin, so that these cells are still responsive to acetylcholine or pilocarpine. Ocular signs of botulism include *pupillary dilatation,* paralysis of accommodation, ptosis and diplopia. Yet the pupils will respond to pilocarpine.

Bromides (potassium bromide and sodium bromide) in overdosage can produce *large pupils* that have a subnormal reaction to light and accommodation. Other features include stupor, psychosis, slurred speech and transitory hallucinations. If the patient recovers, the pupils return to normal.

Cannabis (marihuana, Indian hemp, hashish) can produce central nervous system stimulation with resultant *mydriasis* along with visual hallucinations and disturbances in color vision (ianthinopsia or violet vision).

Coniine (2-propylpiperdine) and *conhydrine* (2-alpha-hydroxypropyl piperidine) are poisonous alkaloids derived from hemlock (*Conium maculatum*), which will produce *mydriasis* along with ascending skeletal muscle paralysis.

Cytisine is an alkaloid derived from *Cytisus laburnum* or *Laburnum anagyroides.* Eating the pods and or seeds of this plant can result in vomiting, collapse and *mydriasis.*

Datura stramonium (Jimson weed, stink weed, thorn apple) has a high content of atropine and scopolamine, and if eaten will produce *mydriasis*. These plants are also used in Christmas decorations and may be eaten by children.

Gelsemium sempervirens (yellow jasmine) contains alkaloids capable of producing *mydriasis* as well as paralysis of accommodation. Oculomotor palsies, especially of the lateral recti, have accompanied severe systemic reactions, i.e. slow pulse, hypothermia and generalized weakness. Diplopia and ptosis may occur as well.

Nutmeg poisoning can produce *mydriasis*.

Oleander (Nerium oleander) is a plant which contains the glycoside oleandrin. Oleandrin, in high doses, produces *mydriasis* in addition to vertigo, convulsions, coma and bradycardia.

Paraldehyde in small doses produces a *miosis* comparable to that found in sleep. However, overdoses of paraldehyde evoke *mydriasis*.

AUTONOMIC EFFECTOR AGENTS[*]

Parasympatholytic agents (anticholinergic drugs). The parasympatholytic agents block the action of acetylcholine at synaptic junctions (ganglia) as well as at nerve terminal sites (effector-cell level), i.e. atropine.

Parasympathomimetic agents. These drugs can be divided into the following:

(a) *Cholinergic* agents, which directly simulate the action of the parasympathetic nervous system (miosis), i.e. pilocarpine. In many instances these agents have a molecular structure that is very similar to acetylcholine. (b) *Anticholinesterase* agents, which act indirectly by neutralizing cholinesterase, an enzyme responsible for the orderly degradation of acetylcholine at synaptic junctions and at nerve-terminal sites (effector-cell level), i.e. eserine.

Sympatholytic. The sympatholytic agents antagonize the action of the sympathetic nervous system, i.e. ergotamine.

Sympathomimetic. The sympathomimetic agents simulate the action of the sympathetic nervous system (mydriasis) by exerting an Adrenalin-like effect. These agents may also potentiate sympathetic nervous system activity by preserving the norepinephrine from enzymatic degradation by monoamine oxidase (MAO) and or catechol-o-methyl transferase (COMT).

I. Parasympatholytic agents.
   A. Natural alkaloids.
      1. Atropine.
      2. Hyoscyamine.

------

[*] Groups I-IV as listed in the outline are derived from Duke-Elder, W. S.: *System of Ophthalmology*, London, Henry Kimpton, 1962, Vol. VII, pp. 535-601.

        3. Scopolamine (hyoscine).
   B. Synthetic cholinergic depressants.
       1. Homatropine (hydrobromide).
       2. Atropine Methyl Nitrate (Eumydrine).
       3. Eucatropine Hydrochloride (Euphthalmine).
       4. Pamine (Scopolamine Methyl Bromide®).
       5. Lachesine (benzylic ester of hydroxyethyl-dimethyl-ethyl ammonium chloride).
       6. Cyclopentolate Hydrochloride (Cyclogyl).
       7. Oxyphenonium Bromide (Antrenyl®).
       8. Amprotropine Phosphate (Syntropan).
       9. Methantheline Bromide (Banthine®).
     10. BL 139 (alpha, alpha-diphenyl-gamma-dimethyl-amino-valeramide).
     11. Dibutoline (di-N-butyl carbamylcholine).
     12. Labotropin (bi-alkyl-aryl-amino ester).
     13. Penthienate bromide.
     14. Pentamethonium.
     15. Hexamethonium.
     16. Tetraethyl ammonium chloride (TEAC), which has no mydriatric effect.
     17. Azamethonium (Pendiomid®), a pentamethonium compound.
II. Parasympathomimetic agents.
   A. Cholinergic agents.
       1. Natural alkaloids.
          a. Pilocarpine nitrate.
          b. Muscarine (alpha-aldehyde-beta-ethylcholine).
          c. Arecoline (methyl tetrahydromethyl nicotinate).
       2. Synthetic cholinergic agents.
          a. Acetylcholine.
          b. Methacholine (Mecholyl or acetyl-beta-methyl choline).
          c. Carbachol (Doryl or Carbamylcholine Chloride®).
          d. Bethanechol chloride (Urecholine®).
          e. Furethonium iodide (Furmethide®).

B. Anticholinesterase agents.
   1. Physostigmine (Eserine)
   2. Neostigmine (Prostigmin)
   3. Organic phosphate esters
      a. Di-isopropyl fluorophosphate (DFP)
      b. Tetraethyl pyrophosphate (TEPP)
      c. Hexamethyl tetraphosphate (HEPT, HEPP)
      d. Diethyl-p-nitrophenyl phosphate (TS 219)
      e. Diethyl-p-nitrophenyl thiophosphate (DPP, DNTP)
      f. Tetra-isopropyl pyrophosphate (TIPP)
      g. Echotiophate iodide (Phospholine Iodide®)
      h. Diethyl-chlormethyl-courmarinyl-phosphoro-thionate (Coralox®)
III. Sympathomimetic agents.
   A. Natural hormones and alkaloids.
      1. Adrenaline (epinephrine).
      2. Noradrenaline (norepinephrine, Arterenol®)
      3. Ephedrine.
   B. Synthetic adrenergic agents.
      1. Phenylephrine Hydrochloride (Neo-Synephrine).
      2. Synephrine (Sympatol®).
      3. Isopropylarterenol (Aludrine®).
      4. Kephrine (Adrenalone®).
      5. Amphetamine (Benzedrine).
      6. Hydroxyamphetamine Hydrobromide (Paredrine®).
      7. Methylaminopropyl Phenol (Paredrenol®).
      8. Naphazoline Hydrochloride (Privine®).
      9. Tetrahydrozoline Imidazoline (Tyzanol®).
   C. Sympathetic sensitizers.
      1. Cocaine hydrochloride.
      2. Procaine (Novocaine®), which has no effect on the pupil when administered via the topical route. However, as part of retrobulbar anesthesia there is mydriasis along with ocular hypotony. Procaine also has some anticholinesterase activity.
IV. Sympatholytic agents.

   A. Natural alkaloids.
      1. Ergot.
      2. Yohimbine (causes a miosis followed by mydriasis).
      3. Corynanthine.
   B. Synthetic sympatholytic agents.
      1. Ergotamine tartrate (Gynergen®).
      2. Dihydroergotamine (DHE), which has no effect on the pupil.
      3. Hydergine (mixture of dihydroergocornine, dihydroergocristine, and dihydroergocryptine).
      4. Dibenzyl chlorethylamine (dibenamine hydrochloride) produces a slight miosis.
      5. Phenoxybenzamine hydrochloride (Dibenzylene®).
      6. Hydrallazine hydrochloride (Apresoline®).
      7. Tolazoline hydrochloride (Priscoline).
      8. Opilon (6-acetoxy-thymoxy-ethyldimethylamine), which produces ptosis and paralytic miosis.
 V. Cellular effectors.
   A. Nitrites.
      1. Amyl nitrite in large doses can produce mydriasis.
      2. Glyceryl trinitrate (nitroglycerin), same as amyl nitrite.
   B. Histamines.
      1. Beta-imidazolylethylamine acts as a miotic and vasodilator.
      2. Aminoglaucosan is a synthetic histamine which acts as a miotic and vasodilator.
   C. Antihistamines, which counter the effects of the histamines.
      1. Antazoline hydrochloride (Antisine®).
      2. Methpyrilene hydrochloride (Histadyl®).
      3. Diphenhydramine hydrochloride (Benadryl®).
      4. Promethazine (Phenergan®).
      5. Pyrilamine (Neo-antergan®).

DEVELOPMENTAL ANOMALIES
OF THE IRIS

I. Hypoplasia of the iris.
   A. Aniridia (irideremia) where there is a rudimentary iris.
   B. Simple coloboma (iridoschisma).
      1. Total coloboma extending from pupil to ciliary body.
      2. Partial coloboma.
         a. Notch in pupillary margin.
         b. Hole in substance of iris (pseudopolycoria, iris dehiscence).
         c. Defect at ciliary border (iridodiastasis).
      3. Complete colobomas. Anomalies that involve entire thickness of the iris.
      4. Incomplete (Pseudocolobomas). Anomalies that only involve part of the entire thickness of the iris.
         a. Mesodermal layer defect (Bridge coloboma). A defect in the stroma and pigment layers.
         b. Ectodermal defect (notches).
         c. Stromal defect (pits in the iris) in the lamina iridopupillaris, the normal crypt or crypts of iris.
   C. Rarefaction of the anterior leaf. Probably represents the recession of the iridopupillary membrane.
   D. Hypoplasia of both mesodermal layers.
   E. Circumpapillary aplasia. Also affects ectodermal structures as well as both mesodermal layers.
   F. Stromal iris atrophy secondary to lues.
II. Hyperplasia of the anterior layers of the iris stroma.
   A. Accessory iris membrane.
   B. Fibrous membrane on the iris.
   C. Splitting of the iris (iridoschisis).

III. Anomalies of the iris musculature.
 A. Congenital microcoria (congenital miosis). Absence of dilator muscle.
 B. Congenital anomalies of sphincter muscle.
  1. Absence of sphincter muscle (circumpapillary aplasia).
  2. Hyperplasia of sphincter muscle.

ANOMALIES OF THE PUPIL

I.   Congenital anisocoria (2% of population). It is an irregular autosomal dominant trait). A difference of 20 percent in the diameters of the pupils in a person is considered significant.

II.   Polycoria.

III.   Dyscoria. Anomalies in the shape of the pupil other than colobomas.
    A. Slit-shaped pupil.
    B. Hourglass-shaped pupil.
    C. Rectangular pupil.
    D. Pear-shaped pupil.

IV.   Corectopia (anomalies in the position of the pupil) or ectopia pupillae. This is inherited as a dominant trait and is often associated with ectopia lentis.

V.   Flocculi. Grape-like nodules that are darkly pigmented. They are found growing from the pigmented epithelium of the iris at the pupillary border. May be seen as a congenital anomaly. They are frequently bilateral and located on the upper pupillary border.

OPHTHALMIC SYNDROMES WITH ALTERED PUPILS

Adie's (Tonic) Pupil. The lesion appears to be in the ciliary ganglion, resulting in a slightly enlarged pupil that is sensitive to 2.5% Mecholyl on local conjunctival instillation. This syndrome is seen at 20 to 40 years of age, mostly in females (75%), and there is often an associated loss of deep tendon reflexes (knee and ankle jerks are depressed). A more detailed account of the clinical findings can be found on page 86.

Argyll Robertson Pupil. The lesion appears to be between the intercalated neurons and the Edinger-Westphal nucleus, within the mesencephalon, resulting in a miotic pupil (see Fig. 26). The pupil does not react to light stimuli. The miosis of the near reflex is unimpaired. This syndrome is usually indicative of neurosyphilis; however there are exceptions (p. 79). Please see main text for detailed discussion of this syndrome (p. 75).

Babinski-Nageotte Syndrome (medullary tegmental paralysis). The lesion appears to affect the descending sympathetic fibers in the pontobulbar region with resultant miosis (ipsilateral). Ptosis, enophthalmos and nystagmus complete the ocular picture. Other clinical features of this syndrome include contralateral hemiparesis with ipsilateral cerebellar ataxia, dysdiadochokinesia and dysmetria.

Benedikt's (Tegmental) Syndrome. The lesion is in the mesencephalon where the third nerve passes either through or next to the red nucleus on its way out of the brain stem. Because of the oculomotor nerve damage, one would expect mydriasis. However, sympathetic fibers are also damaged in this lesion and may make the mydriasis less obvious. There is also contralateral paresis of the extremities as well as intention tremor.

Ipsilateral ataxia of the extremities is also noted as well as the ipsilateral oculomotor palsy.

Carotid-Artery-Venous-Sinus Fistula Syndrome. The lesion is within the cavernous sinus. Pupillary defects will depend upon the damage to the sympathetic and parasympathetic fibers travelling in the cavernous sinus. If both fiber types are damaged, you may not pick up any significant pupillary changes on routine clinical testing. In addition, there is progressive exophthalmos, varying degrees of ophthalmoplegia, secondary glaucoma, dilation of retinal veins, papilledema and eventual optic atrophy. Patient hears a buzzing noise or water running inside his head.

Cestan-Chenais Syndrome. The lesion is located within the lateral portion of the medulla and is due to thrombosis of a vertebral artery. Clinical findings include ipsilateral enophthalmos, nystagmus, ptosis and miosis.

Charcot-Marie-Tooth (Progressive Neuritic Muscular Atrophy). Degeneration of the muscular branches of the peroneal nerves is the initial pathology with subsequent degeneration of all motor neurons supplying the extremities. This is a sex-linked hereditary pattern found in males. The disease usually commences between the ages of 5 and 15 years. Ocular signs include nystagmus and decreased vision. Pupillary disturbances have been reported but are rare.

Dejerine-Klumpke (Lower Radicular or Klumpke's) **Syndrome.** The lesion involves the inferior roots of the brachial plexus, and often there is injury to the cervical sympathetics, giving rise to a Horner's syndrome. Miosis is present, along with ptosis and questionable enophthalmos. In the upper extremities, there is paralysis and atrophy of the small muscles of the hand.

Espildora-Luque (Ophthalmic-Sylvian) **Syndrome.** There is embolic occlusion of the ophthalmic artery which is accompanied by a secondary spasm of the middle cerebral artery. This results in unilateral blindness and a transient contralateral hemiplegia. The pupils are equal but will have altered reactions to the light reflex (ipsilateral).

Feer's (Swift-Feer or Infantile Acrodynia) **Syndrome.** The etiology of this syndrome is unknown, yet the lesions appear to attack the involuntary portions of the central nervous system. The age of onset varies from 1 to 4 years. Ocular findings include proptosis, excessive lacrimation and dilated pupils, photophobia and severe pruritus of the conjunctiva. Other clinical features include inability to sleep, profuse sweating, tremors, tachycardia, hypertension with cyanosis of fingers, etc. Most of these features come under the banner of autonomic nervous system dysfunction.

Fisher's (Ophthalmoplegia-ataxia-areflexia) **Syndrome.** This is a transient polyneuritis whose etiology is unknown. Ocular signs include internal (mydriasis) ophthalmoplegia as well as external ophthalmoplegia. There is also a loss of the deep tendon reflexes.

Foix (Cavernous Sinus, Hypophyseal-Sphenoidal) **Syndrome.** The essential pathology consists of a tumor or aneurysm pressing on the third, fourth and sixth cranial nerves while in the cavernous sinus. Optic atrophy and pupillary mydriasis may be seen. Please see page 84 for discussion of cavernous sinus lesions.

Frenkel's (Ocular Contusion) **Syndrome.** This is blunt trauma to the anterior segment of the globe which results in mydriasis, iridoplegia, hyphema, subluxation of the lens and iridodialysis.

Gänsslen's (Familial Hemolytic Icterus) **Syndrome.** This is a hereditary disease found among Caucasians who show a decreased osmotic resistance or increased fragility of their erythrocytes. In addition to hemolytic crises, polydactyly and splenomegaly, they have iris colobomas.

Guillain-Barré (Landry's Paralysis, Acute Idiopathic Polyneuritis). The etiology is unknown, but there is a chromatolysis of anterior horn cells and cells of Clark's column. Increased albumin in the cerebrospinal fluid has been reported. The oculomotor and facial cranial nerves may also be affected. Fixed, mydriatic pupils have been observed on occasion. Tendon reflexes are absent, along with variable degrees of muscle paralysis usually commencing in the lower extremities.

Horner's Syndrome. The lesion can be anywhere in the long

sympathetic pathway from the posterior and lateral hypothalamic nuclei, within the central nervous system, to the
orbital region involving the postganglionic sympathetic fibers
of the peripheral nervous system. There is ipsilateral miosis,
ptosis, narrowing of the palpebral fissue, etc. Refer to page 89
for a detailed discussion of Horner's syndrome.

Laurence-Moon-Bardet-Biedl Syndrome. This syndrome has a
mixed hereditary pattern affecting mostly Caucasian males.
Ocular findings include abnormal rod receptors within the
retina. There are altered ERG patterns, night blindness,
retinitis pigmentosa and iris colobomas that comprise this
syndrome. Other clinical features include polydactyly, obesity,
mental retardation as well as defects in the cardiovascular
and genitourinary tract systems.

Lowe's (Oculo-cerebro-renal) **Syndrome.** This is an anterior
cleavage or mesodermal dysgenesis syndrome where malformations of the angle are present. Schlemm's canal may be
missing. Congenital glaucoma and nystagmus are often present.
The pupils are dilated and there is no pupillary reaction to
light. Other clinical features include mental retardation and
osteomalacia. There is decreased renal excretion of ammonia.

Marchesani's (Dystrophia Mesodermalis Congenita Hyperplastica)
Syndrome. This is a dominant hereditary disease with defective
closure of the fetal fissure. The lens has excessive curvature to it
(spherophakia) which produces myopia. Many times the
lenses are dislocated posteriorly. When dislocation of the lens
takes place there is iridodonesis. Other features of the disease
include brevocollis, brachydactyly and short stature.

Marfan's (Hypoplastic Form of Dystrophia Mesodermalis Congenita) **Syndrome.** This is an autosomal dominant hereditary
disease with elongation of the metacarpia, and metatarsia.
There is a host of ocular abnormalities, a few of which are as
follows: divergent strabismus, cycloplegia, myopia and colobomas of the optic nerve and uveal tract including the iris.
Dislocation of the lens posteriorly and superonasally is seen
with concomitant iridodensis and possible secondary glaucoma.
Other clinical features include congenital heart disease, dis-

secting aneurysms of the aorta, arachnodactyly and a high, arched palate.

Marinesco-Sjogren Syndrome. This is inherited as an autosomal recessive trait with aniridia and congenital cataracts, as well as cerebellar ataxia and oligophrenia.

Morgagni's Syndrome. There is a dominant inheritance pattern to this syndrome which is seen in females. Bony protrusions into the optic canal is the basic ocular pathology leading to field loss and a possible Gunn pupil. Hyperostosis of the tabula interna of the frontal bone, obesity, hirsutism, headaches, vertigo and tinnitus are also present.

Naffziger's (Scalenus Anticus) **Syndrome.** There are multiple causes of this syndrome, i.e. spinal cord tumors, Pancoast tumor, ruptured nucleus pulposus etc., which consists of compression of the brachial plexus and the subclavian artery by the scalenus anticus muscle. Ipsilateral ptosis, weakness of hand grip and reduced biceps reflexes are present. A miotic pupil (ipsilateral) is often present. Ipsilateral loss of the ciliospinal reflex is also noted.

Nothnagel's (Ophthalmoplegia-Cerebellar Ataxia) **Syndrome.** This is a lesion of the superior cerebellar peduncle, red nucleus and exiting oculomotor fibers. This may be due to a pineal tumor or vascular disturbance of the vermis cerebelli. Oculomotor paresis with gaze paralysis is present, along with a variable amount of internal ophthalmoplegia (mydriasis) and cerebellar ataxia.

Ophthalmoplegic Migraine Syndrome. This is usually a transient vascular problem whose etiology is still unclear. There is severe unilateral supraorbital pain along with transitory homolateral oculomotor paralysis. Internal ophthalmoplegia may occur with mydriasis. The abducens nerve may also be involved.

Pancoast's (Superior Pulmonary Sulcus) **Syndrome.** About 50 percent of cases are due to bronchogenic carcinoma at the apex of the lung (see Fig. 34). Cervical rib erosion is common. Ipsilateral ptosis and miosis (Horner's sndrome) are present along with ipsilateral shoulder pain and parasthesias. Atrophy of hand musculature may take place. See page 89.

Parinaud's Syndrome. Pineal tumors, midbrain lesions, lesions of the white commissure of the pons etc. can cause this syndrome which consists of ptosis, paralysis of upward gaze and retraction nystagmus on upward gaze attempts. Diplopia, displaced pupils and papilledema, as well as vertigo, are present. On occasion, there may be mydriatic pupils (see p. 73).

Parkinson's Syndrome. This is seen in the late stages of epidemic encephalitis as well as in manganese and carbon monoxide poisoning. Neuropathology data reveal widespread destruction of the pigmented cells in the substantia nigra. Ocular manifestations include fluttering of the eyelids, nystagmus, paralysis of convergence and accommodation and pupillary disorders, i.e. mydriasis or anisocoria. "Cog wheel" rigidity of the arms, loss of facial expression, rhythmical tremors and shuffling gait are concomitant features of the syndrome.

Raeder's (Paratrigeminal Paralysis) Syndrome. This lesion involves the sympathetic fibers about the carotid artery as well as those in the fifth (trigeminal) cranial nerve. Frequently a meningioma or an aneurysm of the internal carotid artery can produce this syndrome which consists of the following; mild ptosis, epiphoria, hypotonia and miosis (Horner's syndrome) all ipsilaterally. In addition, there is ipsilateral facial pain with some weakness of the jaw muscles.

Rollet's (Orbital Apex-Sphenoidal) Syndrome. The basic problem in this syndrome is a lesion in the apex of the orbit. There are numerous etiologic agents, including neoplasms, inflammations and hemorrhages, that are responsible. Clinically, there are varying degrees of exophthalmos, ptosis, altered sensation of upper lid, visual field defects (optic atrophy, neuritis, papilledema). The pupil may be average size and fixed, or miotic (see p. 84).

Spheno-Cavernous Syndrome. Lesions of the cavernous sinus which give a picture very similar to that of the superior orbital fissure syndrome (see p. 84). The pupils are semidilated and fixed.

Sylvian Aqueduct Syndrome. This is a tumor or inflammation in the region of the sylvian aqueduct or the third and fourth ventricles or of the superior colliculi. There is retraction nystagmus on upward gaze, as well as extraocular muscle palsies and pupillary disturbances.

Trisomy-D (13-15 Trisomy) Syndrome. There is an extra chromosome in the "D-Group." Ocular findings include anophthalmia, microphthalmia, shallow orbits and hypertelorism. Iris colobomas, cataracts and optic nerve colobomas, as well as numerous other anomalies (polydactyly, dextrocardia, horizontal palm creases, cleft palate, spina bifida, etc.) are part of the clinical picture. Also, cartilage is found on pathologic examination of the globes.

von Herrenschwand's (Sympathetic Heterochromia) **Syndrome.** This is similar to Horner's syndrome (see p. 89). There is unilateral heterochromia with miosis, ptosis and decreased facial sweating, all ipsilaterally.

Wallenberg's (Dorsolateral Medullary, Lateral Bulbar) **Syndrome.** The basic pathology consists of occlusion of the posterior inferior cerebellar artery. This syndrome is seen in patients over 40 years of age. Clinically, there is ptosis and spontaneous coarse nystagmus which are ipsilateral. Diplopia, as well as ipsilateral miosis, is usually present. Other clinical features include nausea, vertigo, ipsilateral ataxia and dysphagia, along with ipsilateral loss of pain and temperature sensation on the face. There is contralateral loss of pain and temperature sensation in the extremities and the trunk.

Weber's (Weber-Gubler, Cerebellar Peduncle) **Syndrome.** This consists of a lesion of the cerebral peduncle, pons or medulla, usually by hemorrhage or thrombosis. There is ptosis, oculomotor palsy and a fixed dilated pupil, all ipsilaterally. Other clinical signs include contralateral hemiplegia in the extremities as well as paralysis of the face and tongue.

INDEX

A

Accommodation-convergence reaction, 51-52
Acetylcholine, 34, 38, 116
Acetylcholinesterase, 36, 38
Aconite, 113
Adie's pupil, *see* Tonic pupil syndrome
Afferent arc, *see* Pupillary pathways
Alkaloids, 118
Amaurosis fugax, 57
Anhydrosis, *see* Horner's syndrome
  skin patterns of, 94T
Anisocoria, 107, 121
  alternating contraction, 73, 74 fig.
Ansa subclavia, 18, 20 fig.
Anterior limiting layer, 11
Anterolateral funiculus, 28
Anticholinesterase agents, 40, 117
Antihistamines, 118
Apnea, 41
Argyll Robertson pupil, 75, 122
  clinical features of, 77-78
  differential diagnosis of, 79
  lesion sites of, 75, 76 fig.
Atrophy, *see* Iris, atrophy
Autonomic nervous system, *see*
    Parasympathetic nervous system;
    Sympathetic nervous system
  effector agents, 115-118
Axon reflex, 54

B

Babinski-Nageotte syndrome, 122
Barracuda meat poisoning, 113
Bell's phenomenon, 52
Benedikt's syndrome, 122
Bibliography, 105-106
Blood supply, *see* Iris, blood supply
Botulinum toxin, 41, 82, 113

Brachium, *see* Superior brachium
Bromides, 113
Bruch's membrane, 25 fig.
Bruch's posterior membrane of iris, 11, 14-15
Budge, ciliospinal center of, 11

C

Calcium, 42
Cannabis, 113
Carbachol, 40, 116
Caroticotympanic nerves, 19, 21 fig.
Carotid artery plexus
  external, 21 fig, 24 fig.
  internal, 21 fig., 24 fig.
Carotid-artery-venous-sinus fistula
    syndrome, 123
Catechol-o-methyl transferase, 36, 115
Cavernous sinus, 19
  plexus, 19
  thrombosis, 84
Cellular effector agents, 118
Central syndrome, 102
Cerebrovascular accidents, 103T
Cervical ganglia, 18
Ceston-Chenais syndrome, 123
Charcot-Marie-Tooth syndrome, 123
Chiasm, disorders, 67
  clinical signs of, 69
  differential diagnosis of, 68-69
Cholinergic agents, 38, 39 fig., 116
Ciliary crypts, *see* Iris, crypts of
Ciliary ganglion, 31 fig., 32-33
  lesions of, 86, 87 fig.
  long sensory root of, 32
  motor root of, 30
  short ciliary nerves, 21
  stimulating agents, 38
Ciliospinal reflex, *see* Pupillary reflexes
Circadian cycles, 17